Facing the Unexpected

Disaster Preparedness and
Response in the United States

Kathleen J. Tierney
Michael K. Lindell
Ronald W. Perry
Editors

JOSEPH HENRY PRESS
Washington, D.C.

JOSEPH HENRY PRESS • 2101 Constitution Avenue, N.W. • Washington, D.C. 20418

The Joseph Henry Press, an imprint of the National Academy Press, was created with the goal of making books on science, technology, and health more widely available to professionals and the public. Joseph Henry was one of the founders of the National Academy of Sciences and a leader of early American science.

Library of Congress Cataloging-in-Publication Data

Tierney, Kathleen J.
    Facing the unexpected : disaster preparedness and response in the United States / Kathleen J. Tierney, Michael K. Lindell, Ronald W. Perry.
      p. cm.
    Includes bibliographical references (p.) and index.
    ISBN 0-309-06999-8
    1. Emergency management—United States. 2. Disaster relief—United States. I. Lindell, Michael K. II. Perry, Ronald W. III. Title.

HV551.3 T54 2001
363.34′8′0973—dc21   2001003004

With thanks and admiration, this volume is dedicated to the four people whose vision and lifelong contributions made it possible:

William A. Anderson, Russell R. Dynes, E. L. Quarantelli, and Gilbert F. White.

Acknowledgments

T his book was written as part of the second national assessment of natural hazards research, a five-year project funded by the National Science Foundation (NSF) under Grant No. 93-12647, with supporting contributions from the Federal Emergency Management Agency, the U.S. Environmental Protection Agency, the U.S. Forest Service, and the U.S. Geological Survey. Dennis S. Mileti of the Natural Hazards Research and Applications Information Center, University of Colorado at Boulder, was principal investigator and overall director for that project. For the second assessment, more than 100 researchers and practitioners conducted a systematic survey of what is currently known and what still remains to be learned on disaster-related topics, ranging from pre-disaster mitigation and insurance protection through post-disaster response and recovery.

Facing the Unexpected is very much a collective product. First and foremost, it reflects the contributions of the other researchers and practitioners who took part in the assessment's Subgroup on Preparedness and Response. Those other group members were Donald Binder, Edward Hecker, Jack Kartez, Jane

Kushma, Janet McDonnell, David Neal, Eric Noji, Brenda Phillips, and Kenneth Stroech. Over a period of more than three years the group collected and synthesized research material, exchanged information, and critiqued chapter drafts. Without their hard work this book would never have been written. We also thank Betty Hearn Morrow, Walter Peacock, Henry Quarantelli, and Robert Stallings for their generosity in reviewing earlier versions of the book and for their very thoughtful suggestions and comments. Thanks are also due to James Makris and Kern Wilson, who reviewed a very preliminary draft of this volume for a 1996 Natural Hazards Workshop session on preparedness and response research. Any errors and shortcomings in this volume are, of course, our fault, and contributors and reviewers should not be held responsible.

A group of graduate students at the Disaster Research Center provided invaluable assistance by searching for articles, preparing bibliographies, and assisting in other ways with the preparation of this volume. Those students were James Dahlhamer, Nicole Dash, Melvin D'Souza, Lisa Reshaur, Tricia Wachtendorf, and Gary Webb. DRC librarian Susan Castelli did a marvelous job of locating materials when we needed them. Keith Appleby prepared charts and figures for the volume, and Rachel Smedley and Rory Connell provided help with the compilation of the final bibliography. Special thanks are due to DRC's office coordinator Mary Ann Brown, whose patience and attention to detail helped get this book into print.

This book is dedicated to four individuals who have contributed immeasurably to the body of research that is reflected in this volume and the other work undertaken as part of the second assessment. Russ Dynes, Henry Quarantelli, and Gilbert White began defining the field of hazards and disaster research more than 40 years ago. They trained and influenced three generations of scholars and practitioners, and they continue to be astonishingly productive scholars and true role models for all of us. Through his leadership at NSF and his commitment to nurturing those successive generations, Bill Anderson has made it possible for the field to grow and flourish. Thank you, Bill, for working tirelessly to keep the societal aspects of hazards and disasters on the research and public policy agenda. That the second assessment was such a major endeavor—that is, that there was so much research to review and evaluate after 25 years—is due in no small measure to your efforts.

Authors

KATHLEEN J. TIERNEY is Professor of Sociology and Director of the Disaster Research Center at the University of Delaware. Her research focuses on how households, organizations, and communities cope with hazards and disasters. She regularly teaches courses on the sociology of disasters, collective behavior, social movements, and qualitative research methods. She is a member of the American Sociological Association, the Earthquake Engineering Research Institute, the International Research Committee on Disasters, and the Executive Committee of the Multidisciplinary Center for Earthquake Engineering Research.

MICHAEL K. LINDELL is Director of the Hazard Reduction and Recovery Center at Texas A&M University, where he currently oversees hurricane hazard analysis and planning for the Texas gulf coast. Since launching his career in environmental hazards through his participation in the First Assessment of Research on Natural Hazards, he has provided technical assistance on radiological emergency preparedness for the International Atomic Energy Agency, the U.S. Nuclear Regulatory Commission, the Department of Energy, and nuclear utilities. He has also worked on hazardous materials emergency preparedness with state emergency response commissions, local emergency planning committees, and chemical companies.

RONALD W. PERRY is Professor of Public Affairs at Arizona State University. He has studied natural and technological disaster management for over thirty years, focusing on floods, volcanic eruptions, earthquakes, and hazardous materials emergencies. Since 1998 he has served as editor of the Elsevier-Science book series, *Contemporary Studies in Applied Behavioral Science.*

Contents

Facing the
Unexpected

CHAPTER ONE

Conceptualizing Disasters and Their Impacts

THIS BOOK IS ONE in a series of volumes that survey and assess research on hazards and disasters. It is part of a large-scale project that reviews both the state of the art and the state of practice in the fields of disaster research and hazards management. That larger project is a follow-up to a landmark assessment of research and applications undertaken more than 25 years ago by Professor Gilbert White and his collaborators at the University of Colorado. The findings from the first assessment were reported in a series of publications that included both major summary volumes (White, 1974; White and Haas, 1975) and more specialized monographs (e.g., Mileti, 1975a, 1975b; Cochrane, 1975; Mileti, Drabek, and Haas, 1975). That same approach is being used in the second assessment project. Reports and books produced by other researchers taking part in the assessment focus on the adoption and implementation of hazard adjustments (Lindell et al., 1997), the role of insurance in providing protection against hazards (Kunreuther and Roth, 1998), land-use planning as a strategy for containing disaster losses (Burby, 1998), and geographic information systems (GIS) as a tool for analyzing hazard vulnerability

1

and disaster impacts (Cutter, 2001). Dennis Mileti's book, *Disasters by Design: A Reassessment of Natural Hazards in the United States* (1999), is an overview volume that condenses findings from the second assessment and advances a perspective on the management of hazards as guided by principles of sustainability.

Nearly three very eventful decades have passed since the first comprehensive assessment was conducted, and the volume of research findings compiled since then more than justifies a new effort to take stock, not only of what we know, but also of what needs further study. The second assessment project is comprehensive, surveying research on all phases of the hazards cycle and on a broad array of topics, including hazard analysis; factors in the societal environment that influence disaster losses; land use planning and management; engineering issues, such as the use of codes and standards in hazard management; disaster predictions, forecasts, and warning; insurance; and disaster recovery.

This book focuses on research that has been conducted on two key topics in the disaster field: pre-disaster planning and post-disaster emergency response activities. Since the time of the first assessment, a large body of research has developed addressing these two subjects. Considerable progress has been made not only in describing and analyzing the preparedness and response activities engaged in by various social units, but also in synthesizing what was already known, developing new theoretical approaches, and methodologically advancing the study of preparedness- and response-related issues.

This research has been driven in large measure by severely damaging and disruptive disaster events that have both intrigued disaster specialists and captured the public's attention. Among those events were the 1979 Three Mile Island nuclear plant accident; the eruption of the Mt. St. Helens volcano in 1980; the Bhopal explosion in 1984; the Mexico City earthquake of 1985; the 1986 Chernobyl nuclear disaster; the Loma Prieta and Northridge earthquakes, which occurred in California in 1989 and 1994, respectively; the 1988 Armenian earthquake; Hurricane Hugo and the Exxon oil spill in 1989; Hurricane Andrew in 1992; the 1993 Midwest floods; and the 1995 Kobe earthquake in Japan.

Since the time of the first assessment of research on natural hazards, periodic flooding in Bangladesh has claimed hundreds of thousands of victims, the 1976 Tangshan earthquake killed an estimated 240,000 people, and many thousands have died in earthquakes and volcanic eruptions in Central and South America. Disaster losses have continued to escalate, both in the U.S. and worldwide. In the last decade, the U.S.

experienced its most costly disaster—the 1994 Northridge earthquake—and losses from that event are currently estimated at $33 billion, a total that is still climbing. That earthquake capped several years of increasing losses in which each major disaster seemed to do more damage and cost more than the previous one. Exactly one year after the Northridge earthquake, the Kobe earthquake killed over 6,000, injured approximately 30,000, and left 320,000 people homeless out of a population of 1.5 million in the impact area. That event caused over $120 billion in losses.

Dramatic disaster events like these have in some cases led to changes in the ways in which disasters and hazards are managed. The Three Mile Island emergency was a significant factor in stemming the trend toward reliance on nuclear power in the United States and establishing detailed standards for evaluating emergency preparedness. The Bhopal disaster had a major influence on U.S. legislation affecting preparedness for chemical emergencies. New federal oil spill management legislation was enacted as a direct result of the Exxon oil spill, and the problems that developed with the emergency response following Hurricane Andrew stimulated efforts to assess and overhaul the federal government's emergency management system. And societal concern about ballooning disaster losses that are increasingly seen as unaffordable has led to a new emphasis on mitigating future damage and on making hazard insurance a more effective loss reduction tool.

Like disasters and their losses, the amount of research available on disaster- and hazard-related topics has increased markedly since the time of the first assessment. Although this body of work has not always been consistent or cumulative, we do know more about a wider range of issues than ever before. This is particularly true with respect to emergency preparedness and response, because most of the research conducted to date has concentrated on topics in those two areas.

The compilation of research findings presented in this volume builds upon previous research on disaster preparedness and response. Summaries and syntheses of earlier research on these topics include work by Dynes (1970); Mileti, Drabek, and Haas (1975); Quarantelli and Dynes (1977); Kreps (1984); and Drabek (1986). In many cases, the more recent research discussed here supports findings from earlier studies, reinforcing what was already known a generation ago. In other cases, however, findings from the classic literature on disasters have been qualified or called into question. In addition to providing insights into long-standing questions in the field, research has also raised new issues that had previously not been considered and has suggested many new topics that warrant study. This volume will review those research findings, point out areas in which knowledge is solid and strong, and identify weaknesses and gaps in the literature. To accomplish these goals we will provide an overview of findings from a wide variety of studies on emergency preparedness and response activities undertaken by households, businesses, community groups, and governmental organizations. In addition, we will discuss the social, economic, political, and cultural factors that shape emergency preparedness and response, as well as the broad societal trends that have influenced disaster management policies and practices in the United States.

THE HAZARD CYCLE AND DISASTERS

Disasters originate in the fact that all societies regularly face geophysical, climatological, and technological events that reveal their physical and social vulnerabilities. In response, societies engage in activities and develop technologies that are designed to provide protection from such threats. However, such measures often prove ineffective and can themselves become a source of added vulnerability when extreme events

Hurricane Floyd left the downtown section of Franklin, Virginia, under six feet of water in September 1999. The water finally receded, but remaining hazards included propane tanks, gas tanks, chemical barrels, and pesticides.

occur. For example, the structures in which we live and work can become agents of death, injury, and damage when wind, water, or ground shaking cause them to fail. Levees, flood channels, and other public works that were originally built to protect communities from flooding can leave them even more vulnerable to large or unexpected floods that exceed their design standards. Similarly, policies and plans designed to provide protection against some types of emergencies may do little to reduce vulnerability resulting from other threats. Even those who survive initial disaster impacts may subsequently become disaster victims should those events result in widespread social and economic disruption.

Disasters are the defining events in a hazard cycle that commonly is characterized by its four temporal stages: mitigation, preparedness, response and recovery (National Governors' Association, 1979). Hazard mitigation involves actions taken before a disaster to decrease vulnerability, primarily through measures that reduce casualties and exposure to damage and disruption or that provide passive protection during disaster impact. Mitigation measures include land-use regulations that reduce hazard exposure and building codes and construction practices designed to ensure that structures resist the physical impacts created by hazards, such as wind, water, or seismic forces. Emergency preparedness encompasses actions undertaken before disaster impact that enable social units to respond actively when disaster does strike. Organizational preparedness activities include developing emergency response plans, training employees and response personnel on what to do in an emergency situation, acquiring needed equipment, supplies, and materials, and conducting drills and exercises. Household preparedness activities include developing an emergency plan for the household, storing food and water, making sure there is a battery-powered radio on hand, and taking other steps to anticipate whatever problems a disaster might create.

Emergency response consists of actions taken a short period prior to, during, and after disaster impact to reduce casualties, damage, and disruption and to respond to the immediate needs of disaster victims. These measures include detecting threats, disseminating warnings, evacuating threatened populations, searching for and rescuing trapped disaster victims, providing emergency medical care, taking action to contain ongoing threats, and providing emergency food and shelter. Finally, post-disaster recovery comprises actions taken to repair, rebuild, and reconstruct damaged properties and to restore disrupted community social routines and economic activities. Recovery activities typically center on the provision of aid for temporary housing and residential reconstruction, the restoration and reconstruction of public infrastructure and facilities, and the provision of assistance to households and businesses that experienced physical damage and other losses. They also aim at reversing whatever negative effects a disaster may have had on the quality of life in an affected community and on the psychosocial well-being of victims. Depending on the severity of the disaster, recovery may take weeks, months, or years. The recovery period is typically also a time in which new mitigative activities are undertaken or at least considered, marking the beginning of another phase in the cycle.

Disasters have a spatial as well as a temporal dimension. Broadly speaking, a distinction can be made between disasters that result in relatively localized areas of severe damage and disruption and those in which impacts are spread over a wide geographic area. At one extreme are disasters, such as the Oklahoma City bombing, in which severe damage was concentrated among a few city blocks. At the other are events like the record flooding that struck the Midwest in 1993, in which nine states were inundated. Dynes (1970) describes the geography of disaster events as involving a series of concentric zones. At the center is an area of very severe impact, which is surrounded by a fringe area in which there is also significant damage and disruption. Aid passes through adjacent filter zones in order to reach the highest impact areas, and more distant community and regional aid zones that are not directly affected by the disaster act as suppliers of resources.

Disasters produce a range of impacts, which can be characterized as direct, secondary or disaster-induced, and indirect effects. Direct effects include the deaths, injuries, and physical damage and destruction that are caused by the impact of the disaster agent itself. Research has recently begun to emphasize the importance of secondary disaster impacts, such as fires or hazardous materials releases that are triggered by earthquakes and environmental pollution resulting from flooding. These kinds of occurrences can produce significant impacts and losses over and above those caused by the primary disaster agent and can complicate response and recovery efforts. A distinction can also be made between direct and secondary impacts and the indirect losses resulting from disasters. Those losses include "ripple effects" resulting from disruptions in the flow of goods and services, unemployment, business interruption, and declines in levels of economic activity and productivity.

Keeping this range of negative impacts in mind, it is clear that disasters can have adverse consequences for the social and economic well-

Aftermath of the Oklahoma City bombing, April 1995.

being of an entire affected area, including households, businesses, and communities that escape direct damage. For example, when the Port of Kobe was severely disabled as a result of the 1995 earthquake, there was concern that Japan's Kansai region and the nation as a whole would suffer economically and that shippers forced to go to other East Asian ports for cargo-handling might not return even after repairs were made. Much emphasis was placed upon repairing damaged Southern California freeways following the 1994 Northridge earthquake because millions of dollars worth of productivity were being lost daily due to transportation delays that affected the entire Southern California region. As these examples show, because developed industrialized societies are increasingly characterized by interdependence among geographic regions and economic sectors, indirect impacts have the potential to ripple out from areas of direct damage following disasters.

CLASSIC APPROACHES TO DEFINING AND STUDYING DISASTERS

Any discussion of research on preparing for and responding to disasters must begin by considering the ways in which the concept of disaster has been used in the social science literature and the different theoretical approaches that have been employed to study hazards and disasters. Over the last 25 years, there has been considerable discussion and disagreement about what constitute the defining characteristics of a disaster and, paralleling that concern, what the subject matter of the field of disaster studies should be (see, for example, Quarantelli, 1982a, 1985, 1987; Hewitt, 1983; Kreps, 1984; Dynes, 1993). The most widely-cited definition of the term in the social sciences is the one developed by Fritz, who defined disaster as (1961a: 655):

> An event, concentrated in time and space, in which a society, or a relatively self-sufficient subdivision of a society, undergoes severe danger and incurs such losses to its members and physical appurtenances that the social structure is disrupted and the fulfillment of all or some of the essential functions of the society is prevented.

To understand how disasters came to be defined and studied in this way, it is important to take into account how the field of disaster research came into existence and the theoretical assumptions that guided pioneering work in the field. United States disaster research originated in questions that the U.S. military had about maintaining social order in wartime situations—for example, whether community residents would panic when faced with a potential or actual nuclear attack. The focus of that research was on disaster events and their immediate consequences, and the primary interest was in practical and applied issues, not necessarily in theorizing about the social origins of disasters. (For an excellent summary of the field's pioneering empirical work and its implications for theory and research, see Quarantelli, 1987.)

As the Fritz definition illustrates, functionalism or social-systems theory has also had a major though largely unacknowledged influence on U.S. disaster research since the field's inception, and many U.S. disaster studies still reflect that perspective, usually implicitly (Kreps and Drabek, 1996; Bolin, 1998). Research on disasters has proceeded on the assumption that societies and communities are systems organized around necessary social functions that from time to time are disrupted by natural and technological agents. After a crisis period necessitating adaptation by affected social units, the social system readjusts and recovery takes place. In one of the classic functionalist formulations in the field, for example, Allen Barton characterized disaster as a type of collective stress situation in which "many members of a social system fail to receive expected conditions of life from the system" (1969: 38). For Barton, what distinguishes disasters from other types of collective social stress, such as war, is that the sources of disasters are external rather than internal.

The functionalist or systems perspective informed other early efforts to conceptualize disaster response, such as the Disaster Research Center's "demand-capability" model (Dynes, Haas, and Quarantelli, 1967). That formulation, applied initially to organizations experiencing the impact of disaster rather than to entire communities or societies, characterized a disaster as a situation producing great organizational stress. The model argues that this stress occurs because sharp and unanticipated demands exceed the capacity of organizations to respond. Further, those demands, which may be quite unusual for a given organization, threaten central values and thus require immediate action. At the same time, organizational capabilities are insufficient to meet escalating demands, both because of the sheer size of the demand "load" and because the disaster itself has degraded capabilities by affecting the availability of personnel and damaging and disrupting facilities. This unexpected, excessive demand requires organizations to adapt if they are to respond effectively.

Thus, the approach most commonly used by researchers to define situations as disasters is based on the functionalist or systems-focused assumption that disasters involve demands that exceed capabilities: when an extreme event impacts a vulnerable community it creates pressure on that community to prevent adverse impacts on public health, safety, and property (Lindell and Perry, 1992). The demands of a small-scale, slow-onset disaster may be such that affected social units can respond on their own, without assistance from larger institutions such as government. By contrast, a large-scale, rapid-onset disaster is likely to also require a timely and coordinated response by many public and private sector organizations to minimize damage and disruption and restore the community to routine functioning. Such coordinated responses may be problematic both because of the magnitude and unexpected nature of the disaster demands and because the organizations that are required to respond lack sufficient training and practice.

When a routinized response is not possible then coping measures must be improvised. The assumption is that the efficiency and effectiveness of both routine and improvised response activities are facilitated by preparedness actions undertaken at the community, organizational, and household levels. Because societies with complex patterns of organization for routine activities require correspondingly complex patterns of organization for nonroutine events like disasters, an assessment of preparedness and response activities requires an understanding of the complex demands these social units face, the tasks they perform, and the manner in which they mobilize resources (Drabek, 1986; Kreps, 1989, 1991; Lindell and Perry, 1992).

This classic theoretical approach to the study of disasters, which blends functionalism and social systems perspectives and looks at disasters as discrete events, seems to have been adopted not so much as the result of conscious choice on the part of researchers, but rather because of the prominence of systems theory at the time the field was developing and the perspective's compatibility with the research methods that were commonly employed in the field. Reflecting the war-related funding priorities mentioned above, from its earliest days most U.S. disaster research has been organized around case studies of disaster events. The typical approach has been to select disasters for study, identify their consequences, and trace the human and organizational responses to those consequences. This event-oriented, inductive research strategy tends to yield results that take the form of models of determinants and consequences assembled in what is often called a general systems framework (Mileti, 1999). Functionalism provides a ready logic that is compatible with the interpretation of such models.

Following this tendency to focus on specific disaster events, researchers also have emphasized the ways in which disaster agent characteristics can affect preparedness and response activities. The disaster properties that have been most discussed in the literature include speed of onset, length of forewarning, magnitude of the physical processes involved (for example, wind speed, wave force, or Richter magnitude), and the geographical scope and temporal duration of their effects (Dynes, 1970; Kreps, 1989). Additionally, disasters vary in frequency and temporal regularity, as well as in the extent to which they are accompanied by environmental cues (Burton, Kates, and White, 1978), attributes which have been found to have important consequences for both emergency preparedness and response. For example, regularity and speed of onset affect the ability to forecast the location and timing of disaster impact. A longer warning period makes it possible to issue warnings to the public and to increase response capability—for example, by notifying emergency responders of the threat and moving emergency equipment to pre-designated staging areas. Forewarning also allows threatened communities to engage in expedient mitigation actions, such as boarding up windows and tying down objects. We would expect response activities, therefore, to be more effective and losses to be lower in disasters for which warning is possible.

With respect to scope of impact, the expectation is that disasters with community-wide and regional impacts will be more difficult to manage than those in which damage and disruption are more localized. One reason for this is that larger disasters have more of a tendency to disrupt the infrastructure of an affected area, making damage assessment, communication, the movement of resources, and other response-related tasks more difficult. Moreover, in disasters with a large scope of impact there are typically fewer unaffected community residents available to provide assistance to victims. This necessitates the mobilization of emergency aid from other areas, the activation of mutual aid agreements, and participation by state, regional, and federal agencies, thus expanding the need for interorganizational and intergovernmental coordination.

Hazard agents also differ in the extent to which they are familiar to community residents and emergency responders. Familiarity is generally a function of the degree of prior experience a community has had with a particular disaster agent. Of course, experience can lead to both desirable and undesirable outcomes. On the one hand, experience may make particular hazard agents more salient to community residents and local officials, stimulating preparedness and response efforts. On the other, it may engender complacency or fatalism. Additionally, because communities have a tendency to plan for the types of events that are more frequent and thus more familiar, they may neglect less frequent, catastrophic or low-probability/high-consequence events in their planning.

THE NATURAL HAZARDS PERSPECTIVE

Like the classical disaster research approach, the environmental hazards research perspective predated the first assessment. However, in contrast with the functionalist and event-based orientation of the disaster research tradition, natural hazards research views hazard vulnerability as the product of the joint functioning of a natural events system and the human use system (White, 1974; Burton, Kates, and White, 1978; Sorensen and White, 1980). According to this approach, societies occupy physically vulnerable locations in the course of their search for resources such as fertile land, commercially advantageous locations, and even attractive surroundings and scenic views. Disasters occur when the risk area population adopts patterns of land use, building construction, and economic activity that are vulnerable to the physical impacts of extreme events in the physical environment, such as tornadoes or floods. When broadened to include environmental sources generally, rather than only natural hazards, this perspective can also encompass biological hazards, such as crop fungal diseases, and technological hazards, such as radiological materials and toxic chemicals.

According to the hazards model the risk of the undesirable impacts that can result from extreme environmental events can be reduced through the adoption of hazard adjustments. These adjustments may be biological (reducing physiological vulnerability to hazards) or cultural (reducing vulnerability through behavioral adaptation). Cultural adjustments include a range of measures, from abandoning or changing the use of a particular location in order to avoid the impacts of extreme events through modifying those events and taking steps to bear the impacts and share the burdens of their occurrence. The most common adjustments are those that aim at preventing the injurious effects of the hazard agent (Burton, Kates, and White, 1978). This goal is often approached through public works (e.g., dams and levees), through the implementation of warning and evacuation systems, building codes, and hazard-resistant construction practices designed to enhance population protection, and through economic practices such as making appropriate choices for crop selection and planting. However, sharing the burdens resulting from extreme events through the provision of post-disaster relief is also common.

CONCEPTUAL DEBATES IN THE STUDY OF DISASTERS

The functionalist or event-based formulation still serves as the basis for much of the research undertaken on disasters in the United States. Mainstream approaches characterize disasters as suddenly-occurring disruptions, originating from either natural or technological sources, in which the demands placed on the social system to respond exceed the resources or capabilities possessed by that social system. However, classical ways of delimiting the subject matter of disaster research are increasingly being questioned. For example, how "concentrated in time and space" does an event have to be before it is considered a disaster? Are long-term environmental processes like global climate change and desertification by definition not disasters because their onset is slow in human terms? Should chronic threats that suddenly or cumulatively begin to have acute effects be considered disasters? Is AIDS a disaster? What about famine and war? How should we deal with failures in technology that distribute themselves over very wide geographic areas or that occur in cyberspace rather than in the physical world? What if people define a hazardous situation like repeated flooding or ground-water contamination as normal rather than as an emergency and decide to just live with it? Does that mean no emergency exists? Why are the 110 people killed in a Valujet crash in the Florida everglades considered disaster victims, while the 50,000 killed annually in traffic accidents in the U.S. are not? The 1979 accident at the Three Mile Island nuclear plant was a major crisis that occasioned very extensive social disruption in affected communities. However, were those communities and their residents actually in danger when the accident occurred? If so, how severe was that danger? Is disaster an objective phenomenon or a social construction? Is it useful to think of disasters as events at all, or can they more productively be conceptualized as processes? Are disasters occurrences that impinge on social systems from the outside—that is, from the environment that supports the social system—or are they immanent in the social order itself? Should disasters more appropriately be viewed as social problems rather than as discrete events (Drabek, 1989; Kreps and Drabek, 1996)? These are the kinds of issues that often arise in discussions concerning disasters and their effects.

A recent publication entitled *What Is a Disaster?* (Quarantelli, 1998a) was devoted entirely to discussions and critiques of the concept of disaster as used in the social science literature. In addition to raising questions like those above, contributors offered a range of views on the substance and meaning of the term. Gilbert (1998) outlined three paradigms that have been used in the field: the war analogy, which sees disaster as an external agent "attacking" and disrupting the social system; disaster as a manifestation of vulnerabilities inherent in the social order; and disaster as connected to uncertainty—that is, as a disruption of systems of meaning and understandings of cause-effect linkages—which is based in turn on increasing societal complexity. Dombrowsky (1998) argued that disasters occur because human activities, which have both intended and unintended consequences that are not well understood, interact and come into conflict with ongoing technological and natural processes. Frequently, we don't know that human activities and natural processes are on a collision course—or even what might go wrong—until a disaster actually happens.

In that same volume, Kreps (1998) took the position that the Fritz definition should be retained with some modification and that disasters can most productively be conceptualized as events that are sudden and dramatic, that involve social disruption and harm, that generate a collective response, and that at least in theory can be mitigated. At the same time he argued that disasters are social constructions; that is, disaster events and their impacts do not exist *sui generis,* but rather are products of social definition (see also Kreps, 1989). Porfiriev (1998) defined disasters as involving social system destabilization, some degree of destruction, excessive physical and psychological demands, and the necessity for undertaking emergency actions to bring about a return to stability. Dynes (1998) identified several types of disasters, ranging from those in which affected communities can cope with impacts more or less on their own to disasters that involve different forms of community dependency on outside resources. Also included in his conceptualization were events that do not involve entire communities but rather are confined to specific institutional sectors, as well as potential community threats that become the focus of public attention and mobilization. In an earlier article on the same topic, Horlick-Jones (1995) argued in favor of defining disasters as originating in the fundamental social conditions of late-modern society and as involving disruptions of cultural expectations and the release of existential dread. Such dread or anxiety originates in turn in a loss of faith in the institutions that are supposed to keep risks under control.

This lack of consensus on ways of conceptualizing disaster is related to some degree to longstanding issues in the field. One such debate centers on whether disasters should be defined primarily by their physical characteristics and impacts or by their social dimensions. On the one hand, like the Supreme Court justice who can't define pornography but knows it when he sees it, disaster researchers decide what to study by looking for events involving physical damage and bodily harm, which suggests that physical properties are important defining characteristics of the phenomenon. Researchers who argue that natural and technological disasters differ in their impacts clearly rely to some degree on physically-based conceptualizations. Indeed, the past 25 years have seen a major debate on the issue of whether natural and technological disaster agents differ in ways that are significant for our understanding of preparedness and response activities. This debate, which we discuss in more detail in Chapter Six, has been fueled in part by the aftermath of catastrophic events like the Three Mile Island and Chernobyl nuclear accidents, the Bhopal disaster, and the Exxon oil spill, as well as by the conflicts and controversies generated by toxic chemicals and hazardous wastes. One body of research suggests that disasters caused by technological agents constitute a distinct genre because the social and behavioral patterns that occur in emergencies and disasters involving technological agents differ from those that are commonly observed in natural disasters, and because the two types of events tend to differ in their short- and longer-term consequences. Some researchers therefore take the point of view that research findings based on studies of natural disasters cannot be generalized to crises originating in failures of technology.

On the other side of the debate are researchers who argue that, rather than making categorical distinctions between natural and technological disaster agents, it is more important to focus not on the origins of the agents themselves but rather on similarities and differences that cut across the natural/technological distinction—characteristics such as speed of onset, warning period length, and spatial scope of impact. According to this view disaster events with similar characteristics will produce similar behavioral responses and emergency management challenges regardless of whether they originate in the natural environment or in technology. Exemplifying this perspective, Quarantelli has long argued that disasters and their impacts are social rather than physical phenomena and that "we should conceive of disasters for sociological purposes only in social terms" (1989a: 247). This notion would apply not only to the social characteristics and attributes of disaster situations (their social impacts, social factors in disaster vulnerability, organized efforts to respond, and so forth) but also to the social-structural causes or sources of disaster victimization. This latter view—that disasters are social occasions as well as physical events—is central to social scientific disaster research, and it forms the basis for the discussions in the chapters that follow.

In considering these conceptual debates, it is important also to recognize that the appropriateness of any definition can vary, depending upon the purposes for which the definition is being used. In many cases, these purposes are theoretical, and definitional differences reflect disciplinary divisions. Just as physical scientists have conceptualized disasters in terms of their physical dimensions, social scientists have used definitions and measures that are congruent with their own disciplinary back-grounds—psychosocial impacts for psychologists, organizational and community impacts for sociologists, or direct and indirect losses for economists. Thus, from a psychological perspective a disaster is an event that threatens the psychological functioning of its victims, while from an economic perspective it is an event that produces measurable material losses and threatens the flow of goods and services. These discipline-related differences on ways of thinking about what constitutes a disaster have often stood in the way of theoretical and research integration. Many recent arguments about how to conceptualize and define disasters reflect a healthy attempt to break out of the discipline-based approach that has characterized work on disasters and hazards.

For others, definitions of what constitutes a disaster may be rooted in practical, rather than theoretical, concerns. Indeed, practitioners' conceptualizations of disaster may well be broader than those of researchers, encompassing issues that transcend any single discipline. For example, community planners and emergency managers view the physical characteristics of disaster impacts to be important indicators for defining vulnerable areas and see hazard analysis as providing an important basis for decisions regarding land use and construction practices. They may also be interested in using vulnerability analysis to identify groups that have a high probability of becoming victims should a disaster occur and to project victim needs. At the same time, they must be concerned with potential negative economic impacts and with the political aspects of managing hazards and disasters. Thus, while researchers' definitions of disaster have been primarily discipline-based, the perspective of practitioners is interdisciplinary. This difference may be one factor impeding the dialogue between practitioners and researchers.

EMERGING THEORETICAL PERSPECTIVES

In recent years theoretical perspectives other than functionalism have begun to have an impact on how disasters are conceptualized and studied. Among these alternative approaches are social constructionism (see, for example, Stallings, 1991, 1995; Kreps and Drabek, 1996); recent European critiques of modernity and industrial society (Luhmann, 1993; Beck, 1992, 1995a, 1995b; Horlick-Jones, 1995); conflict-based and political-economy theories (Hewitt, 1983; Stallings, 1988; Tierney, 1989; Bolin, 1998); and political-ecological perspectives (Bates and Pelanda, 1994; Peacock, Morrow, and Gladwin, 1997).

The social constructionist approach to disasters, which is related to the constructionist perspective in the social problems literature (Spector and Kitsuse, 1987; Miller and Holstein, 1993; for a wider-ranging use of social-constructionist assumptions see Sarbin and Kitsuse, 1994), argues against viewing disasters as objective physical phenomena with given properties and impacts. Rather, according to the constructionist critique, ways of defining and labeling hazards and disasters—whether an event constitutes a disaster, how probable and potentially damaging disasters are, and what can be done to reduce their impacts—are socially produced through organized claims-making activities. From this perspective, it is not what happens or could happen in the physical world—death, damage, and disruption—that is important for understanding disasters, but rather the social processes through which groups promote claims about disasters and their consequences. For example, Stallings's analysis of the earthquake problem (1995) shows how views on the severity of the earthquake threat and strategies for managing seismic risk were shaped by a small group he calls the "earthquake establishment"—engineers, geologists, and seismologists from universities, the private sector, and government. The social construction of the earthquake problem was channeled not by the concerns of the general public but rather by the institutional interests of scientific disciplines whose work centered on the study of earthquakes and government agencies that were trying to contain the economic losses that could result from a catastrophic earthquake.

Those favoring a constructionist perspective of course do not argue that earthquakes, floods, tornadoes, or other agents of harm do not exist. Rather, they point to the importance of exploring the social activities in which interest groups and different stakeholders engage as they try to place disaster-related problems on the public agenda and elicit the kind of governmental and institutional response they believe is warranted. Constructionists contend further that the processes through which hazards, disasters, and their causes and consequences are socially defined are by no means as straightforward or nonproblematic as mainstream disaster researchers assume. According to this view, the properties of disasters—even such seemingly objective properties as severity or scope of impact—are not inherent in the phenomena themselves but rather are the product of social definition. The fact that such definitions of impact, injury, and loss are not as strongly contested in cases like the Kobe earthquake as they are in cases like Love Canal and Three Mile Island should not obscure that point.

One significant trend in U.S. disaster research appears to be toward a synthesis of functionalist and constructionist perspectives. For example, Kreps and Drabek have argued that disasters can usefully be conceptualized as nonroutine social problems (i.e., as involving the same kinds of claims-making and definitional activity that accompanies the construction of other social problems). Their position is that (1996: 142):

> . . . the essence of disaster is the conjunction of historical conditions and social definitions of physical harm and social disruption at the community or higher levels of analysis. During and immediately following an event, claims-making and response activities translate as domains of collective action to meet demands that are socially defined as acute. A large-scale mobilization takes place to meet these needs, existing groups and organizations restructure existing activities, and new structural forms are socially created.

Thus (1996: 143), "the emergent research questions relate to social definitional processes and the behavioral activities reflective of societal adjustments to hazards."

In a very different vein, European social theorists have recently begun turning their attention to hazards and disasters, especially environmental ones, in ways that highlight what is unique about disasters in the developed world. Beck (1992, 1995a, 1995b) describes present-day industrial society as a "risk society" with distinctive characteristics that include the transformation of formerly calculable risks into massive incalculable threats; the appearance of previously unknown threats, such as nuclear, chemical, and genetic hazards, from which the institutions of society cannot offer protection; and the emergence of institutions whose role it is to symbolically control the uncontrollable, deny the existence of threats, and transform threats into (seemingly more manageable) risks. In the risk society (Beck, 1995a: 2) "[t]hreats are produced industrially, externalized economically, individualized juridically, legitimized scientifically, and minimized politically." Unlike the position taken by classical scholars such as Barton and Fritz, this position sees the potential for disasters as immanent in the social order itself, rather than originating outside it, and conceptualizes disasters as an inevitable and direct consequence of the social relations and practices that characterize modern society.

In *Risk: A Sociological Theory* (1993), Luhmann considers similar issues. Central to his argument is the distinction between risks and dangers. Risks are potential losses that are viewed as the consequences of decisions, while dangers are losses attributable to the environment—that is, losses that aren't perceived as resulting from choice but rather as acts of God or nature. Because of the high degree of "structural coupling" (1993: 98) between the institutions of modern industrial society and technology, society has become riskier. Risk is inherent in technologies, and today's high technologies generate ever larger risks. Further, rather than being reduced risk is intensified through practices that use technology to regulate the safety of technology. Following Luhmann, the disasters at Three Mile Island, Chernobyl, and Bhopal—events that involved not only failures of technology but also failures of technological control systems—are examples of the linkage between disaster potential and the way industrial society is structured. Risk is, in other words, an inherent feature of modern social systems.

Luhmann argues that modern life involves not only higher dependence on decisions—in other words, greater risk—but also an inability to identify what decisions and whose decisions actually produce undesirable outcomes. In earlier times, people made individual decisions about whom to trust and potentially affected parties assumed risks. In modern industrial society, however, rather than making choices individuals and groups find themselves increasingly affected by decisions made by others.

Another challenge to the classical view that disasters originate outside the social system—that is, in the system's environment—comes from scholarship that has been influenced by conflict-oriented perspectives such as critical theory, political economy, and world systems theory. Resembling work that has been done on the environment (e.g., Buttel, 1976; Schnaiberg, 1980; Schnaiberg and Gould, 1994), these conflict-based approaches to the disaster problem view disaster victimization as a consequence of the exercise of political power by elites and of the dynamics of the capitalist world system. In U.S. society, for example, development interests that promote intensive land-use development and economic "growth" are seen as a main source of escalating disaster losses (Tierney, 1992). Globally, so-called underdeveloped countries are vulnerable to environmental extremes because of their incorporation into a world system that keeps them dependent and marginal (Susman, O'Keefe, and Wisner, 1983). Disaster vulnerability is thus inextricably linked to the processes that promote dependency and underdevelopment.

In a related analysis, Hewitt (1983) argued that, contrary to most thinking in the disaster research tradition, disasters do not result from the failure of systems to adapt to environmental extremes but rather are closely interwoven with ongoing social life. Rather than being caused by exceptional environmental processes or extreme environmental events, natural disasters are the "normal" outcomes of particular sociopolitical strategies, formulated to benefit privileged groups, that have as their consequence increased risk to others. A similar criticism of functionalist and event-based approaches to the study of disaster is voiced by Bogard (1988: 154), who argues that the American disaster research tradition has:

> . . . reproduced our commonsense idea of disasters as temporally bounded events in the environment necessitating a response. The content of that response, however, was perceived as constrained by certain essential features of disaster itself, as an event in the environment over which little or no control could be exercised. The unpredictability of disaster, its perceived externality to the routine of social life, its characterization as an "act of God," all entered into and reinforced this idea.

Along these same lines, Bolin has called for an approach to studying hazards that sees disasters and their impact as resulting from political-economic forces that simultaneously shape both the vulnerability of the built environment to disaster damage and the social vulnerability of exposed populations. For Bolin (1998: 9-10):

> Vulnerability concerns the complex of social, economic, and political considerations in which peoples' everyday lives are embedded and that structure the choices and options they have in the face of environmental hazards. The most vulnerable are typically those with the fewest choices, those whose lives are constrained, for example, by discrimination, political powerlessness, physical disability, lack of education and employment, illness, the absence of legal rights, and other historically grounded practices of domination and marginalization.

Unlike mainstream disaster research, a political-economy/conflict perspective sees governments not as "champions" of hazard reduction (Lambright, 1985) but rather as key actors in bringing disasters about, either through the passive acceptance or the outright promotion of hazardous activities. For example, it was the U.S. government that, for strategic and military reasons, promoted the development of nuclear power (Clarke, 1985) and spurred oil exploration in the Alaska wilderness and oil shipping in Prince William Sound (Gramling and Freudenburg, 1992). Viewed in this light, Three Mile Island and the *Exxon Valdez* oil spill were byproducts of the pursuit of power and profit by an industry/government partnership. In the natural hazards area, a political-economy analysis would locate the source of the massive damage and disruption caused by Hurricane Andrew not in the storm's 200-mile-an-hour gusts but rather in the politics of land development in South Florida, the short-term profit orientation of real estate entrepreneurs, and collusive local governments accustomed to looking the other way when good design, engineering, and construction practices were not followed. Similarly, when the government steps in to provide assistance to victims and communities when disaster strikes, that intervention is a continuation of its normal role, which is to ensure the smooth operation of the economic system (Stallings, 1998). (For related discussions, see Clausen, et al., 1978; Dombrowsky, 1987; Stallings, 1988; Tierney, 1989; Bolin, 1998.)

A related body of work, the ecological-vulnerability perspective, also focuses on the economic, political, and social sources of victimization and loss. In *At Risk* (1994), Blaikie and his co-authors develop a framework that characterizes disasters as involving the convergence of socially produced vulnerability and exposure to hazards. Vulnerability to disasters is produced ultimately by political, economic, and ideological/cultural processes that put individuals and groups at risk and by institutions that fail to provide adequate protection. Underlying the increase in disaster vulnerability are interrelated global processes that include (1994: 32) "population growth, rapid urbanization, international financial pressures (especially foreign debt), land degradation, global environmental change, and war." Rather than having random or unpredictable effects, disasters disproportionally harm socially vulnerable groups that have already been marginalized by the class system and by racial, ethnic, gender and other forms of discrimination. Similarly, Oliver-Smith argues for what he terms a "political ecology" approach to disasters centering on "the dynamic relationship between a human population, its socially generated and politically enforced productive and allocative patterns, and its physical environment" (1998: 189).

The emerging ecological-vulnerability perspective sees communities not as unitary systems but rather as consisting of loosely-coupled, heterogeneous ecological elements and networks (Bates and Pelanda, 1994; Peacock, Morrow, and Gladwin, 1997). Within these ecological groupings power and resources are not distributed equally. Rather, relationships among units are shaped by gender, racial, and ethnic stratification, economic inequality, and differential access to political power. These differences in turn influence the ways diverse segments of the community experience and cope with disasters. According to this view, understanding disasters and their impacts thus means taking into account "sociopolitical issues such as the extent to which social inequality, heterogeneity and complexity, competition and conflict, and coordination exist within the network of social systems" (Peacock and Ragsdale, 1997: 27).

LINKING HAZARDS AND DISASTER RESEARCH

By questioning taken-for-granted assumptions about disasters and introducing new theoretical perspectives that sensitize researchers to features of the social order and processes that were previously overlooked, recent scholarship promises to transform the field of disaster studies. One way of better understanding these newer contributions is to explore the linkages that have developed between research influenced by the hazards tradition, on the one hand, and disaster research on the other. As illustrated in Figure 1.1, research in the environmental hazards tradition, which was conducted mainly by geographers and planners, focused principally on understanding hazard vulnerability and on *pre-event* adjustments, mainly the adoption of mitigation and preparedness measures. Hazards research was most concerned with geological, meteorological, and hydrological hazards and paid almost equal attention to urban and rural vulnerability. In contrast, most disaster research, which was conducted primarily by sociologists, focused to some degree on preparedness but mainly on *pre-, trans-,* and *immediate post-impact* response activities and secondarily on disaster recovery. Disaster research addressed a broader range of hazard agents, including both natural and technological hazards as well as "dissensus" community crises such as civil disorders, and it focused on describing and analyzing the activities of various social units, ranging from households to community groups and organizations and governmental authorities at the local, state, and federal levels. And, as noted above, in part because researchers directed their attention to the disaster event as the primary subject for analysis, disaster research has tended to focus on immediate antecedent conditions, response-related behaviors, and the relatively short-term consequences of disasters as opposed to the broader social context in which disasters occur or their longer-term consequences.

As the recent theoretical developments discussed above demonstrate, the differences that previously existed between the hazards and disaster research traditions have broken down as researchers have begun to develop more comprehensive perspectives that consider both disaster events and the broader structural and contextual factors that contribute to disaster victimization and loss. While the functionalist approach that characterized classical disaster research mainly addressed the fact of disaster, not the sources of disaster vulnerability, other work has sought to better understand the societal processes that create vulnerability, how vulnerability is distributed unequally across societies, communities, and social

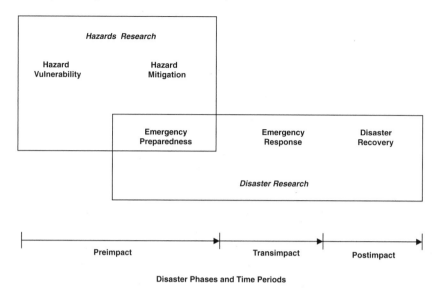

Figure 1.1 Substantive Foci of Hazards and Disaster Research

groups, how vulnerability changes over time, and how and why these changes come about.

In summary, since the time of the first assessment disaster research in the U.S. has moved in the direction of greater theoretical diversity. Broadly speaking, current research is guided by three general theoretical approaches: the functionalist or systems perspective, the vulnerability perspective, and social constructionism. As they have in the past, functionalist assumptions continue to have a significant influence on research, including in particular studies that focus on emergency preparedness and response. Indeed, many scholars (see, for example, Kreps and Drabek, 1996) would contend that, whether as a matter of conscious choice or not, most disaster research reflects functionalist assumptions, in that researchers have sought to understand disaster events in terms of their social-systemic antecedents and consequences. At the same time, American disaster scholarship also incorporates a variety of non-functionalist perspectives that have in common their focus on the economic, political, and social processes that affect disaster vulnerability. Social constructionism provides a third framework for analyzing hazards, disasters, and their impacts that is compatible with both functionalist and vulnerability-focused approaches. This theoretical diversity has sparked debate and stimulated further theoretical refinement. While some researchers have been critical of functionalist theorizing and have proposed alternative theoretical models, others have argued for its continued relevance; still others have looked for ways of bridging and synthesizing different theoretical perspectives. (For a lengthier discussion of theoretical diversity in the field of disaster research, including contributions from non-U.S. disaster theory and research, see Bolin, 1998).

ORGANIZATION OF THIS BOOK

This book begins by discussing work on the topic of disaster preparedness and then moves to consider advances in knowledge in the area of disaster response. Throughout, we attempt to show how the most recent generation of disaster research is related to what we thought we knew 25 years ago, to point out which findings appear at this time to have the most support and where gaps exist in our knowledge, and to suggest ways of addressing important questions raised by this body of work.

The research findings discussed here fall into two general categories. The first consists of empirical studies on specific social units and processes, specific communities, or specific disaster events. Examples of this type of research include studies on how households in a given community prepare for earthquakes, on evacuation following the issuance of disaster warning for a particular disaster event, or on interorganizational emergency preparedness networks in a particular community. The second category of research findings might best be described as reviews or overviews whose conclusions apply across a number of quantitative or qualitative studies. In many cases, works falling into this category attempt to synthesize and generalize from studies whose methodological approaches are dissimilar but whose findings are consistent with one another.

Because disaster research has tended to focus on either preparedness or response—but generally not the two in concert—those two topics are treated separately here. However, the book's organization also recognizes that both sets of activities have been shaped by common societal forces and trends. Following the review of the literature on preparedness and response in Chapters Two, Three, and Four, the social factors and societal characteristics that affect both preparedness activities and post-disaster response are discussed in Chapters Five and Six.

One of the insights derived from the political-economy approach to disasters is that preparedness and response are linked to the development process and related to larger issues of sustainability. Although researchers who have studied preparedness and response have seldom attempted to make that connection, it is possible to situate these activities in a broader sustainability context. In Chapter Seven we discuss disaster preparedness and response, as well as more general questions concerning disaster vulnerability, from the point of view of this newly-developed approach.

We should note at the outset that, with respect to both emergency preparedness and response research, detailed, systematic knowledge declines as the level of analysis moves from smaller to larger social units. For example, we know much more about the preparedness activities of households than we do about those of communities. Similarly, since the time of the first assessment considerably more large-scale research has been done on household response—particularly response to warnings and evacuation recommendations—than on organizational and community response activities.

Additionally, regardless of unit of analysis there are very few areas in which we can claim our knowledge is adequate. Two important objectives of this review, then, are to assess the extent to which we have confidence in our research findings (both the old and the new) and to identify key areas of both theoretical and policy significance where our knowledge is clearly deficient. Throughout the book, we offer suggestions for future research and identify potentially productive directions that research might take.

CHAPTER TWO

Getting Ready: Research on Disaster Preparedness

T HE PROCESSES INVOLVED IN preparing for disasters have been a major research focus since the field of disaster research began. Broadly speaking, the objective of emergency preparedness is to enhance the ability of social units to respond when a disaster occurs. The preparedness process begins with hazard and vulnerability analyses that attempt to anticipate what problems are likely to occur and proceeds with the development of ways to address those problems effectively. The primary goal of emergency preparedness is for households, businesses, and government agencies to develop appropriate strategies for responding when disaster occurs. Preparedness also aims at ensuring that resources necessary to carrying out an effective response are in place prior to the onset of disaster or that they can be obtained promptly when needed. For communities, preparedness encompasses a wide range of activities. These include formulating disaster plans; providing training for disaster responders and the general public to improve their understanding of what to do in a disaster as well as their performance of disaster-related tasks; and conducting emergency response drills and exercises. Other preparedness activities include acquiring equipment, facilities, and other material resources that will enable an effective response when a disaster strikes, and carrying out actions aimed at increasing public hazard awareness. Similarly, for households and organizations preparedness involves being ready to take self-protective actions and being able to obtain the resources needed for both an effective response and recovery.

A large proportion of the emergency preparedness literature focuses on preparedness for natural hazard events. However, one trend that has been very evident since the time of the last assessment is an increasing emphasis on preparedness for disasters involving chemical, nuclear, and other technological hazard agents. This new focus has been driven by a growing awareness of the problems associated with hazardous technologies—events such as the Three Mile Island and Bhopal accidents, and legislation such as the 1986 Superfund Amendment and Reauthorization Act. As a consequence, the literature on preparedness has expanded to consider a much wider range of hazard agents than before. At the same time, the extent to which findings based on studies of technological disaster planning carry over into the natural hazards area, and vice versa, remains unclear. Factors such as methodological differences among studies, the paucity of studies focusing on multiple hazards, variations in the social, economic, and cultural settings in which studies have been carried out, and the fact that studies have been conducted in non-comparable hazard contexts complicate efforts to arrive at generalizable conclusions.

The major activities associated with preparedness, such as planning, emergency drills and exercises, and the stockpiling of emergency supplies and equipment can be engaged in by various social units: households, businesses and governmental agencies, communities, supra-community entities such as states and regions, and entire societies. In this chapter we first review findings from research on preparedness, focusing on successively larger units of analysis. We will begin with households and businesses, continue with government agencies at the local level, and conclude with the state and federal levels. We then consider a series of more generic issues related to preparedness which the literature has attempted to address—often with varying degrees of success. Specifically, we discuss evidence on whether preparedness has an impact on the effectiveness of emergency response activities and on the efficacy of different strategies used to enhance preparedness.

HOUSEHOLD PREPAREDNESS

Encouraging People to Prepare: The Risk Communication Process

Clearly, one of the most significant impediments to enhancing emergency preparedness among households is the low salience of disasters in most people's lives. Members of the public may not receive preparedness information, fail to act or put off taking action, or lack the resources to prepare. As is the case with other pre-impact actions such as insurance purchase (e.g., Kunreuther et al., 1978), decisions about whether and how much to prepare do not conform to the assumptions of classical economic theory. Recent research indicates that while individual decision making contains systematic rational components it also is subject to a variety of heuristics and biases (Feldman and Lindell, 1990) as well as random errors that intervene between the receipt of hazard information and the adoption of hazard adjustments (Palm et al., 1990). Preparedness decisions are influenced by a broad range of factors that cannot adequately be captured using simple rational choice assumptions. Further, both the information people need in order to prepare and the resources they possess in order to carry out preparedness measures are unequally distributed throughout society.

An understanding of how and why households prepare for disasters must be based first on an understanding of how the public perceives and acts on risk information. Recent research on information processing has helped to clarify the ways in which objective hazard levels, hazard perception, and the adoption of preparedness behaviors are related. Nigg (1982) described these stages as hearing the information, understanding it, and perceiving its relevance. An alternative formulation drawn from research on persuasive communications by Lindell and his colleagues (Lindell and Barnes, 1986; Lindell and Perry, 1992) characterizes emergency preparedness decision making as comprising five stages: attention, comprehension, acceptance, retention, and action. A similar typology drawn from cognitive research on information processing identifies six slightly different phases: exposure, attention, encoding, retrieval, judgment, and action.

Based on their research, Mileti and his collaborators (Mileti, Fitzpatrick, and Farhar, 1990; Mileti and Fitzpatrick, 1993) suggest that successful risk communication—that is, communication that stimulates action—is based on four general principles. First, risk communication is a process, and the impact of such communications cannot be understood unless the risk message is placed in context along with other such communications. Second, risk communication involves the joint effects of source and message characteristics (e.g., source credibility, repetition, frequency of repetition, specificity, type and number of channels used to disseminate information) on the one hand and the characteristics of members of target audiences, including their sociodemographic characteristics and experience with the hazard, on the other. Third, risk perception is multidimensional, involving hearing, understanding, believing, and personalizing a risk. Finally, what people do when they receive risk information is the result not only of the information itself but also of other activities in which people subsequently engage, such as evaluating the risk information that has been provided, seeking additional information from other sources, and discussing the risk information with friends, relatives, neighbors, and coworkers.

Obviously there is considerable overlap among these formulations, but they also differ in some respects and the significance of those differences is not always clear. Lindell and Perry (1992), for example, found attention to be an important characteristic in distinguishing among the hazard awareness and education programs catalogued by Sorensen and Mileti (1987). Moreover, the stages of comprehension and acceptance appear to be closely connected with the contents (as opposed to the processes) of cognition. At this point, we have only a rudimentary understanding of the processes by which new information about hazards is integrated with existing risk perceptions, and further work to clarify this process is needed.

Research has documented various ways in which sociodemographic and sociocultural factors affect both the receipt of risk information and what people ultimately do with the information they receive. For example, there appear to be significant differences between the channels community residents use most and those they prefer, as well as variations among both communities and ethnic groups within the same communities in the communication channels that are preferred and used (Perry and Lindell, 1990a, 1990b). For example, with respect to ethnic variations in the receipt of hazard information, Perry and Nelson (1991), questioned samples of whites, African-Americans, and Mexican-Americans regarding preferred channels for receiving information. They reported that ethnic minority groups differ among themselves as well as from the majority group in their reliance on different sources of hazard information. However, all three ethnic groups indicated that information garnered in the past was dominated by the mass media—radio, newspapers, and television—although Mexican-Americans reported obtaining proportionately more information through social networks than either African-Americans or whites. Four distinct patterns were found among the ethnic groups with respect to their preferred modes of information receipt. First, all three ethnic groups had obtained hazard information in the past via radio, and this source remained high on their lists of preferred choices. Second, none of the three ethnic groups identified either speakers at meetings or magazine articles as information sources they preferred. Third, Mexican-Americans were much more likely to list neighborhood meetings as a preferred source than were either of the other groups. Finally, only African-Americans and Mexican-Americans listed television as a preferred source of communication; whites tended to favor written forms of information.

This study was limited in that the minority participants tended to have low incomes, making it impossible to disentangle ethnic from economic effects. Although the research was subsequently replicated in another community with greater income variance (Nelson and Perry, 1991), we still know little about the hazard information sources higher-income minority citizens prefer (Perry, 1987).

Other research reinforces the idea that ethnicity and other social factors affect access to sources of information on hazards. Turner, Nigg, and Heller-Paz (1986) found that both African-Americans and Latinos were more likely than Anglos to depend almost totally on the mass media for information related to earthquake predictions, as opposed to using both the media and other sources, such as books and informal discussions. This is significant, the authors note, because media messages that are not reinforced, confirmed, or corrected through discussion are probably less likely to have an impact on behavior. Older people, the unmarried, people living in households with no school-age children, and the less educated also tended to rely primarily on the media for information about the earthquake threat. In contrast, people with higher incomes and more education supplemented the news media with other information sources.

These kinds of findings are important, because receiving information on why and how to prepare is clearly a precondition for later stages. Variations in household preparedness levels are attributable at least in part to variations in the sources people use for obtaining information, which are related in turn to the content of that information and to the impression it makes. Although some of the studies discussed below address this issue, further research is still needed to determine what channels are likely to be the most effective in reaching different segments of a hazard prone community and encouraging them to prepare.

From research conducted over the last 25 years we know that many impediments prevent authorities from communicating with the public in ways that succeed in getting people to prepare. One of these impediments is uncertainty in the messages that are being conveyed. Scientists often disagree about the probability and likely severity of different threats, and this lack of consensus can cloud risk communication efforts. Problems with source credibility pose additional barriers. Members of the public may be unable to distinguish scientifically qualified sources of hazard information and preparedness advice from less qualified ones or, faced with conflicting information, they may be unable to determine which message is based on good science and which is not. Vivid examples can make more of an impression on people than statistical data. Dramatic disaster events may lead them to be overly concerned about certain hazards while neglecting ongoing threats. Judgments about risk levels can be shaped more by an awareness of the potential consequences of catastrophic events than by their historic frequency of occurrence. Both industrialized and developing societies are characterized by considerable geographic mobility. As people migrate and resettle, they leave familiar areas—and familiar hazards—for less familiar ones. The hazard-related knowledge they gained through living in their former communities may not be relevant in their new environments, and they may be unaware of the new risks they face. Risk information must compete for attention with numerous other types of information that may be much more salient to the public. And even when public information efforts make people more aware of hazards they still may not take the required protective actions (Covello, Slovic, and von Winterfeldt, 1987).

The Nature and Extent of Household Preparedness for Disasters

How much time, effort, and money are people willing to invest in preparing for disasters? Who is most likely to prepare, and why? When people do prepare, which preparedness measures do they favor and why? Researchers have been trying to answer these kinds of questions since the field of disaster research began, and as a result of the large volume of research on household preparedness we now have a much better picture of the preparedness process at the household level and the factors that influence that process.

Tuscaloosa, Alabama, December 20, 2000 — A tornado victim walks out of the makeshift below-ground shelter he built some ten years earlier. The man, his wife, and their two dogs survived an F-4 tornado in that shelter after hearing warning sirens. The tornadoes killed a total of 12 people in Alabama.

One important methodological advance in research on household preparedness over the last 20 years has been the development of emergency preparedness inventories or checklists that provide a systematic way of measuring the kinds of preparedness measures households adopt. Table 2.1 shows one such inventory, which has been used in a number of studies, including research on household responses to the earthquake threat in Southern California (Turner, Nigg, and Heller-Paz, 1986) and to the pseudo-scientific Iben Browning earthquake prediction in the Central United States (Edwards, 1993). These kinds of standardized measures have made it possible to identify which activities are most frequently undertaken, which are preferred by households, and what factors or dimensions are involved in the selection of preparedness strategies. Standardized inventories also make it possible to conduct research comparing household preparedness levels both across communities and in the same community over time. More recently, preparedness inventories developed for use with households have also been adapted for use assessing preparedness among businesses.

TABLE 2.1 Typical Household Earthquake Preparedness
Checklist Items

• Store food or water	• Learn first aid
• Get a first aid kit or medical supplies	• Develop a family emergency plan
• Have a working flashlight	• Have a working battery-operated radio
• Protect dishes and glassware in cabinets	• Strap the water heater to a wall
• Ask about earthquake insurance for your home	• Buy earthquake insurance for your house
• Talk to children about what to do in an earthquake	• Bolt or strap heavy furniture to the walls to keep it from falling over
• Have an engineer or other qualified person assess the safety of your home	• Make structural changes to your home as indicated for safety reasons
• Attend meetings on how to prepare for an earthquake	• Try to get information on how to prepare for an earthquake

Based on Turner, Nigg, and Heller-Paz, 1986.

Despite these advances, findings from household preparedness studies still should be interpreted with caution. Even though there has been movement toward standardizing the measurement of household preparedness, the theoretical variables that have been examined in studies on preparedness, the operationalization of those variables, and the research designs used still vary considerably. Even more important, generalizing from existing studies remains problematic because research on preparedness has been undertaken over a range of different hazard contexts. Some studies, for example, focus on preparedness during normal times—that is, in the absence of a recent disaster occurrence or stepped-up efforts to inform the public in anticipation of a coming event. Others measure preparedness in the context either of recent disasters or of disaster predictions, forecasts, and enhanced public education efforts. While this profusion of approaches has yielded a wealth of new and useful ideas, the idiosyncratic nature of many studies and their linkage to particular hazard contexts leaves a good deal of uncertainty about their replicability and generalizability (Lindell and Perry, 2000).

The rather large body of work on household preparedness for earthquakes exemplifies both the contributions and the limitations of household preparedness research. In one of the earliest studies on this subject Jackson and Mukerjee (1974), surveying San Francisco residents, found that the majority of the sample had experienced an earthquake and most had experienced one or more tremors while living in San Francisco. Nearly half the respondents believed another earthquake would occur in the next few years, and a majority expected an earthquake to affect them personally. However, only about one-third agreed that residents of the city "have trouble" with earthquakes and an even smaller proportion expected to experience significant damage if an earthquake were to occur. Almost half the respondents were unaware of any measures they could take to reduce earthquake damage. While approximately half the respondents considered structural adjustments to their homes and the purchase of insurance a good idea, only about 7 percent had actually done these things. These results were again substantiated by Jackson (1977), who found that 69 percent of the respondents in his California sample had taken no precautionary measures to reduce seismic hazards. This research found that adoption of preparedness measures was associated with previous earthquake losses.

In the same area and at about the same time, Sullivan, Mustart, and Galehouse (1977) conducted research in 1970 and 1976 among residents living along the San Andreas Fault in San Mateo County, California. The investigators found that nearly 80 percent of the respondents were aware that they lived a mile or less from the fault. Most knew of the fault's location before moving there and indicated that they would feel no safer if they lived five miles further from it. The only adjustment measure addressed by the investigators—insurance purchase—increased from 5 percent in 1970 to 22 percent in 1976, an effect the investigators attributed to the 1971 San Fernando earthquake.

An extensive examination of factors affecting the purchase of earthquake insurance by Kunreuther and his colleagues (1978) found that many homeowners in risk-prone areas lacked information on the earthquake hazard. Compared with those who purchased insurance, the uninsured tended to consider an earthquake less likely and to expect lower property damage from a severe event. Perhaps most significantly, the researchers found that one-fourth of the uninsured didn't know that insurance was available and that those who did know were not able to estimate the cost of coverage or had inflated estimates of how much it would cost.

Several years later, Turner, Nigg, and Heller-Paz (1986) reported on an extensive study they had conducted on the public response to the earthquake hazard in Southern California. The project focused on a number of topics, including the ways in which households responded to scientific and non-scientific reports of earthquake precursors like the Southern California Uplift (also known as the Palmdale Bulge) and to small earthquakes that could be interpreted as indicators that larger ones might occur. Five sets of interviews were conducted between January, 1977, and December, 1978, with each wave including questions about the salience of the earthquake hazard, attitudes toward earthquake prediction, the public's understanding of phenomena like the Uplift, preparedness actions undertaken in anticipation of an earthquake and in response to small temblors, and judgments about how well government was prepared for such an event.

Respondents initially showed high levels of awareness of the earthquake hazard, but this was largely due to pseudo-scientific predictions and general earthquake forecasts rather than to their knowledge of scientific predictions. Over time, all sources of information tended to have less of an impact on residents' attention and recall.

Earthquake preparedness was found to be significantly related to the level of hazard awareness, with those who had heard, understood, and personalized the risk being much more likely than those who had not heard about potential earthquake precursors like the Uplift. Nevertheless, the majority of those surveyed had undertaken no preparedness measures at all. The researchers found that levels of preparedness were related to recent experience with a damaging earthquake and personal contact with friends, relatives, and others who were trying to prepare for earthquakes. Various measures of community attachment—having school-age children, being married, owning a home, and having lived longer in the community—were also found to have a positive impact on preparedness levels.

Using a similar design, Dooley and his colleagues (1992) studied earthquake concerns among residents of Orange County, California, at six-month intervals over a three-year period. They found that concern about the earthquake problem rose immediately after each of two significant earthquakes, but in both cases had declined again by the time of the next survey. Concern about earthquakes was found to be positively related to levels of household preparedness. Consistent with the findings in the Turner, Nigg, and Heller-Paz study, levels of preparedness were higher for respondents who were married, had children in the household, were older, and had lived longer in their current homes.

In a study of household preparedness in Memphis and Shelby County, Tennessee, in 1990 at the time of the pseudoscientific Iben Browning Central U.S. earthquake "prediction," Edwards (1993) found that awareness of the prediction was virtually universal. Belief in the prediction was also found to be relatively high: 44 percent of respondents thought a damaging earthquake was likely in the Memphis area within the time-frame covered by the Browning forecast. The survey found that residents were extensively involved in seeking and sharing information about earthquakes but that awareness, concern, and information-seeking didn't automatically translate into action. Asked about a range of things that could be done to prepare for earthquakes, most of the Memphis respondents had taken only about half of the recommended precautions, only 14 percent had undertaken more than half of those measures, and 9 percent had done nothing at all.

The Edwards study replicated many of the findings reported by Turner, Nigg, and Heller-Paz, including the existence of positive relationships between earthquake preparedness levels and the presence of children in the home, educational levels, and household income. These associations can be explained in terms of parents' attentiveness to children's safety, greater ability to understand complex information on the hazard among the more educated, and higher levels of disposable income, some of which can be used to better prepare the household for possible disasters. Like the Los Angeles investigators, who found that ethnicity was related to the propensity to prepare, Edwards found that whites were more likely than African-Americans to engage in preparedness activities.

In another study on the household response to the Browning prediction, Showalter (1993) found high levels of awareness of the prediction, moderate levels of belief, and moderate reported involvement in preparedness activities. Of those responding to her survey, just under 30 percent had attended public meetings to obtain more information about the earthquake hazard, and 20 percent reported making physical changes to their homes to reduce potential earthquake damage. At the same time, 16 percent indicated that they had not done anything to plan for a coming earthquake and that they did not intend to do so.

In research conducted in a different hazard context, Mileti and O'Brien (1992) studied preparedness levels and their association with aftershock warnings that were issued following the Loma Prieta earthquake, which struck the San Francisco Bay Area in 1989. Their study of 734 San Francisco and 918 Santa Cruz County residents found that most were aware of the aftershock warnings, particularly in Santa Cruz County, and that a majority of the respondents (66 percent in San Francisco County and 75 percent in Santa Cruz County) believed that damaging aftershocks would occur. Two months after the earthquake, substantial numbers of people had taken one or more additional preparedness measures, such as protecting household items from damage. Preparedness was generally higher for Santa Cruz County residents.

Significantly, Mileti and O'Brien found that the people who were most likely to pay attention to and act on aftershock warnings were those who had already experienced damage in the Loma Prieta event and who subsequently got involved in the emergency response. People who were not affected by the mainshock tended to do less in response to aftershock warnings, leading the researchers to hypothesize that (Mileti and O'Brien, 1992: 53):

> Those who experience little or no loss in the impact of a disaster may be prone to a 'normalization bias' when interpreting post-impact warnings for subsequent risk: 'the first impact did not affect me negatively, therefore subsequent impacts will also avoid me.'

Mileti and Fitzpatrick (1993) studied community residents' responses to the public information campaign that accompanied the U.S. Geological Survey's Parkfield, California, earthquake prediction "experiment." The Parkfield experiment constituted yet another hazard context: a credible scientific prediction and ongoing monitoring project on a segment of the San Andreas Fault that was expected to produce a significant earthquake in the near future. The information dissemination strategy developed to encourage the public to get ready for a Parkfield event involved the distribution of printed material on the earthquake hazard to local residents. To assess the impact of this public education effort, Mileti and his colleagues sent surveys to household samples in three communities (Coalinga, Paso Robles, and Taft) that varied in distance from the earthquake fault and in recent earthquake experience. The following were among their findings: awareness of both the hazard and the prediction experiment was high; the public awareness campaign had led residents to personalize the earthquake risk; there was some increase in levels of household preparedness, generally involving actions that were easier and less costly to undertake; and proximity to the fault and recent earthquake experience heightened both public awareness and levels of preparedness.

The work conducted by Mileti and his collaborators on the Parkfield prediction found generally that people were more likely to remember the prediction, understand and believe it, consider themselves to be personally at risk, and take protective action if they (1) saw the risk communications they had received as consistent with one another; (2) remembered details of the earthquake prediction, such as projected magnitude and damage potential; (3) remembered specific guidance they had been given to protect themselves against earthquake damage; (4) recalled receiving risk communications through several different channels; and (5) perceived the hazard information as having come from many respected information sources, such as official sources and scientists as well as from relatives and other informal information sources.

To a large extent the study confirmed what is already known about hazard communication—for example, that before acting on information provided by official and media sources people search for additional information and interact informally to confirm reports they have received. However, the authors also highlighted the importance of printed material in communicating moderate-term risk (as opposed to short-term warning), arguing that materials like the household brochure that was used in the Parkfield information campaign, which people can keep and use as a reference as needed, have a greater impact on knowledge and behavior than more ephemeral forms of communication. (For other findings and practical implications of this research see Mileti, Fitzpatrick, and Farhar, 1990; Mileti and Fitzpatrick, 1993.)

In a subsequent study (and yet another hazard context), Mileti and Darlington (1995) conducted research to assess the cumulative effects of the Loma Prieta earthquake, widespread media coverage of seismic hazards, and the dissemination of a detailed newspaper insert on the earthquake hazard to San Francisco Bay Area residents. The researchers found that a substantial segment of the local population expected an earthquake to strike the area in the next few years but were generally optimistic about avoiding personal loss. Respondents were generally well-prepared for earthquakes; many preparedness actions had already been taken prior to the distribution of the newspaper brochure, but levels of preparedness also rose in the year following its dissemination, indicating it may have had some impact. For example, the proportion of respondents who reported stockpiling food and water rose from 44 percent to 75 percent, and earthquake insurance purchases increased from 27 percent to 40 percent.

In a more social-psychological vein, Mulilis and his colleagues (Mulilis and Lippa, 1990; Mulilis and Duval, 1995) conducted a series of studies examining the usefulness of protection motivation theory in predicting earthquake hazard adjustments. Mulilis and Lippa (1990) distributed specially-prepared earthquake awareness brochures to 111 homeowners in Orange County, California. The brochures systematically varied information about an earthquake's probability of occurrence, its severity, the efficacy of a recommended hazard adjustment, and the receiver's self-efficacy or capability to implement the adjustment. The specialized information brochures did induce short-term changes in receivers' perceptions of earthquake probability and severity and of outcome efficacy and self-efficacy, but these impacts were not sustained over the period between the first and second post-tests.

In related research, Mulilis and Duval (1995) tested the proposition that adoption of earthquake adjustments is a function of residents' appraisals of their personal resources (self-efficacy and response efficacy) relative to the demands associated with a threatening event (probability, severity, and imminence), arguing that those who appraise their resources as sufficient are more likely to engage in problem-focused coping strategies such as the adoption of preparedness measures, rather than using emotion-focused strategies. Specialized brochures that varied information about earthquake threats and personal resources produced corresponding differences in respondents' perceptions on these two dimensions. Respondents also differed systematically in their adoption of earthquake hazard reduction measures over the subsequent month. Those who perceived their resources as significantly greater than the demands associated with the event were more likely to prepare than those who saw their resources as equal to or significantly less than what the event would require.

Russell, Goltz, and Bourque (1995) compared data collected between 1988 and 1990 on household preparedness in Los Angeles County and the San Francisco Bay Area with findings from similar studies that had been conducted during the 1970s. Their data included preparedness measures collected both before and after damaging earthquakes struck those regions. (The earthquakes were the Whittier Narrows event, which struck greater Los Angeles in October of 1987 and the Loma Prieta earthquake, which did damage throughout the Bay Region in October of 1989.) The study is noteworthy because the survey items used to measure preparedness closely resembled those used in the Turner, Nigg, and Heller-Paz Southern California survey (1986), which had been conducted about a decade earlier. (That study, as noted above, was conducted in the context of growing concern with the earthquake hazard, stimulated by presumed earthquake precursors like the Southern California Uplift.)

The Russell, Goltz, and Bourque study found that prior to the two earthquakes levels of household preparedness had improved in both Los Angeles County and the Bay Area, but only along one dimension. Households were more likely than before to take survival-oriented precautions, such as keeping supplies on hand and learning first aid than they had been ten years before. However, they were actually less likely to engage in some planning activities, such as developing a household disaster plan, than they had been ten years earlier. Similarly, they were no more likely to have taken steps to mitigate earthquake damage, such as installing earthquake-resistant latches on cupboards or rearranging their shelves. Slight improvements were seen in preparedness in both study areas following the two earthquakes, but these changes were not dramatic. The study also showed that while households were taking many of the steps that are recommended in order to prepare for earthquakes, particularly survival-oriented ones, only a very small number of the measures asked about in the survey were undertaken by more than half the respondents.

The factors associated with higher levels of preparedness before the two earthquakes were home ownership, higher income and educational levels, being married, the number of children at home, the length of time living in the neighborhood, and the extent of previous earthquake experience. Besides indicating that financially better-off residents have a greater propensity to prepare, the findings also suggest that community attachment is a factor. Post-earthquake preparedness levels were affected by proximity to the earthquake's epicenter, the amount of damage households experienced in the earthquake, pre-earthquake preparedness, and other more psychological variables, such as how much fear respondents reported experiencing at the time of the earthquake and how much they continued to think about their earthquake experiences. Moreover, households that had done more to prepare prior to the earthquake took fewer steps after it occurred. Russell, Goltz, and Bourque (1995) suggested that this may be either because there were fewer things left to do (since they had already undertaken a number of preparedness measures), or because in their judgment what had already been done was sufficient.

More recently, Lindell and Perry (2000) have concluded from their review of the literature on seismic hazard mitigation and emergency preparedness that risk-area residents' perceptions of the characteristics of different hazard adjustments are likely to significantly affect their intentions to adopt these adjustments, as well as their subsequent behavior with respect to preparedness measures. This idea was borne out in later research by Lindell and Whitney (2000). Specifically, their data showed that respondent's ratings of the efficacy of different preparedness measures in protecting persons and property, as well as their perceived utility for other purposes, were highly correlated with both intentions to adopt those measures and reports of actually having adopted them. In contrast, requirements with respect to funds, knowledge, skill, time, effort, and cooperation from others appear to exert a less significant influence. Moreover, consistent with research conducted by Mulilis and Duval (1995), respondents' perceptions of their personal knowledge about the hazard as well as their personal responsibility for taking action were also significantly associated with intentions to adopt and with actual adoption of hazard adjustments.

A related study conducted by Lindell and Prater (2000) examined the mitigation and preparedness activities undertaken by residents in both high (Southern California) and low (Western Washington) seismic hazard areas. They found that Southern California residents had much higher levels of personal hazard experience and moderately higher levels of hazard intrusiveness, a measure of how frequently people reported thinking about, discussing, and receiving information about earthquakes. Nevertheless, Southern Californians had only modestly higher levels of risk perception, defined in this study as the perceived likelihood of personal injury and property damage. Moreover, differences among residents in the two risk areas in the adoption of mitigation and preparedness measures were trivial. Income and marital status were found to be the only two of ten different demographic variables to significantly predict the adoption of hazard adjustments, and both hazard experience and hazard intrusiveness were found to be more strongly related to adoption than was risk perception.

Moving next to research involving other hazards, Perry and Lindell (1990a, 1990b, 1990c) focused on hazard awareness and preparedness among residents of the area around Mt. St. Helens in the three years after the volcano's May, 1980, eruption. Comparing responses of residents in two different communities—one 25 miles immediately downstream from the volcano and the other 45 miles downstream—they found that residents of the community closer to the volcano were able to name a larger number of possible consequences of the volcano threat than were those who lived further away. Factors associated with higher levels of knowledge included the amount of damage experienced in the 1980 eruption, volcano-related employment, presence of school-aged children in the home, and frequency of contact with authorities. The number of hazard consequences noted by respondents was also associated with respondents' reports of their perceived risk to personal safety and property, the salience of the volcano hazard, overall levels of personal planning activity, and the actual number of preparedness measures they had adopted. Significantly, the number of hazard consequences respondents could recall was a stronger predictor of levels of preparedness than were ratings of perceived vulnerability or reports of past damage. Such knowledge was not related to age or income.

Similar results were obtained from a study conducted near the Mount Shasta volcano. Adoption of protective measures was strongly related to perception of risk, the presence of children in the household was positively correlated with information-seeking, and even though there were no visible signs of volcanicity awareness of the volcanic hazard was high (Perry, 1990). The positive correlation between perceived risk and both knowledge of the threat and adoption of protective measures were also confirmed in related research focusing on the Mount Usu volcano threat in Japan (Perry and Hirose, 1991).

Faupel, Kelley, and Petee (1992) explored the question of whether disaster education programs (including specific programs centering on the earthquake hazard) affected the extent to which households in South Carolina had prepared prior to Hurricane Hugo. Preparedness was conceptualized as involving planning activities and other adaptive behaviors, such as storing food and having a battery-operated radio on hand. General disaster education did have a positive impact on preparedness levels; however, earthquake-specific educational activities did not carry over to affect hurricane preparedness levels. Among the other factors found to be related to some aspect of household preparedness were prior hurricane experience, having dependents in the home, and home ownership. The study found that whites were more likely to prepare than African-Americans and members of other minority groups. However, education and income levels, which other studies have found to be important predictors, did not predict preparedness levels in the South Carolina sample.

Factors Associated with Household Preparedness

Research conducted to date suggests that people are encouraged to prepare for disasters under three conditions. First, the threat of disaster must be seen as high in the short-term—as occurs, for example, when a specific warning or hazard advisory has been issued for a given community. Second, the source disseminating the hazard and preparedness information must be seen as credible. And third, the preparedness information must be provided repeatedly through different channels and in a form that is easy to recall and use (e.g., in a printed brochure).

It appears to be difficult to stimulate household preparedness for any hazard when people believe there is a low probability of a near-term threat. Why this is the case is not clear. Members of the public may pay less attention to preparedness messages under those circumstances, and consequently remember less, or they may tend to focus more on emotion-centered coping responses such as denial of the threat during such periods. It is also possible that they attend to, comprehend, and accept preparedness messages but postpone action until later. A further troubling finding from Mileti and O'Brien's work on response to aftershock warnings is that even when a damaging disaster has recently occurred, households that escaped damage may subsequently have a tendency to disregard messages about an ongoing threat. These findings are consistent with Kates's (1962) earlier characterization of people as "prisoners of their experience." Evidently many people have a tendency to believe that what already *has* happened is the worst that *can* happen.

As a result of research undertaken in the past 25 years we know that household preparedness activities are socially structured, and we have a much clearer idea of the social factors that influence household preparedness. Actions to protect the household are more likely to be undertaken by those who: are routinely more attentive to the media (primarily those who are educated, female, and white); are more concerned about other types of social and environmental threats; have personally experienced disaster damage; are responsible for the safety of school-age children; are linked with the community through long-term residence, home ownership, or high levels of social involvement; and can afford to take the necessary steps to prepare.

Indeed, one of the most important contributions of the research conducted over the last 20 years has been to examine the importance of socioeconomic factors in household preparedness decisions. The literature suggests that, other things being equal, households with higher socioeconomic status are better prepared for disasters than their financially less-well-off counterparts and that ethnic minorities show a lower propensity to engage in emergency preparedness activities. People who are poor and marginalized have fewer resources to devote to preparedness and have less access to information on hazard reduction (Perry and Mushkatel, 1984; Turner, Nigg, and Heller-Paz, 1986). What is not well understood is how strategies can be developed to reach under-informed and under-prepared populations and to make preparedness feasible and affordable for the financially less-well-off.

Moreover, what we have learned to date on household preparedness suggests that even households that are attempting to address preparedness issues are doing relatively little. For example, Lindell and Prater

(2000) found that the average number of emergency preparedness activities reported by their respondents was slightly less than eight out of a possible 16, and that there were only slight differences in preparedness between residents of highly vulnerable Southern California, where the mean number of measures adopted was 8.3, and less vulnerable Western Washington, where the average was 7.4. Affluent, better-educated white homeowners may be more likely to prepare for disasters, but even their levels of preparedness tend to be low compared to what they could be doing.

Additionally, while we may have some degree of understanding of the factors that are correlated with differing levels of preparedness, we are only just beginning to understand the social-psychological processes involved in the adoption of self-protective measures. In other words, we know quite a bit about *who* prepares but not *why* they do so. Both the intention to prepare and actual preparedness behaviors appear to be related to personal knowledge about hazards, perceived personal responsibility for taking action, and perceptions about the characteristics of different hazard adjustments (Lindell and Whitney, 2000; Lindell and Perry, 2000).

Many of the studies discussed above have focused on public awareness campaigns undertaken to improve household preparedness in the context of growing awareness about presumed near-term threats, such as the Parkfield earthquake prediction (Mileti and Fitzpatrick, 1993). Others, such as the work of Mileti and his colleagues (Mileti and Darlington, 1997) which focuses on efforts to educate the public in the San Francisco Bay Area through the distribution of a brochure on the earthquake hazard, followed closely upon large disaster events. Questions remain about the extent to which findings from these kinds of studies can be generalized to non-threat situations or to geographic areas lacking recent disaster experience. Further research is needed to better understand what motivates people to increase and sustain preparedness efforts during periods of relative normalcy.

Finally, we want to re-emphasize the point made at the beginning of this section. While the volume of information on household emergency preparedness has grown tremendously in the last 25 years, large gaps exist in our knowledge. Moreover, the research cannot be considered cumulative because so few attempts have been made to replicate previous findings. The fact that much of the work undertaken has been done in very different hazard contexts, ranging from situations involving hazard exposure but no recent disaster history to situations involving a high short-term likelihood of damage (e.g., the danger of earthquake after-shocks) limits the generalizability of what has been found. Similarly, we must be cautious in generalizing findings from research in settings where there has been a concerted effort to reach the public and encourage them to prepare to other situations where such strategies have not been used.

ORGANIZATIONAL PREPAREDNESS

This section discusses research on preparedness that uses the organization as the unit of analysis or that focuses on the preparedness measures adopted by particular types of organizations. Studies that describe or analyze preparedness at the interorganizational and community levels, including studies of general and specialized interorganizational networks, are discussed in the next part of this chapter.

Like research on households, knowledge concerning organizational preparedness and the factors that encourage organizations to prepare is still quite uneven. Considerably more is known about preparedness activities among public sector organizations—particularly local emergency management agencies and other crisis-oriented organizations—than about other types of organizations. However, even this research is far from comprehensive. Further, although a number of studies address general preparedness issues, researchers have also tended to focus on organizational preparedness for specific kinds of hazards, such as chemical emergencies or accidents involving hazardous wastes (Sorensen and Rogers, 1988; Faupel and Bailey, 1988).

Clearly, the same factors that constrain preparedness at the household level also exist at the organizational level. Hazards have low salience for most organizations except when there is an imminent threat, and potential disaster-related problems must compete with other more pressing concerns on an organization's agenda. Moreover, organizations that are experiencing financial difficulty will tend to downplay preparedness if it is seen as low-priority or optional; and even when a danger is recognized the resources necessary to deal with it may not be adequate. Generalizing from research in the broader literature on implementation, Waugh (1988) has argued that preparedness programs are difficult to implement because of five general types of impediments: the overall intractability of the disaster problem; the lack of clear and measurable performance objectives; insufficient resources; inadequate levels of public and official support; and the fact that higher governmental levels provide insufficient emergency management expertise and guidance to local communities.

From a practical point of view, emergency preparedness is a central concern for only a very small number of organizations. For the large majority, disaster-related issues are peripheral or incidental to organizational goals and priorities. It follows that the less an organization sees itself as having important disaster functions the more difficult it will be to stimulate preparedness.

This difference in the priority placed on disasters—that is, the distinction between organizations for which responding to disasters is a primary organizational mission and those for which it is not—will serve as an organizing device for the discussion that follows. We begin by looking at preparedness activities among crisis-relevant organizations and then move on to consider other kinds of organizations, including private businesses.

Crisis-Relevant Organizations: Local Emergency Management Agencies

The preparedness activities engaged in by local emergency management agencies have been a major research emphasis since the disaster research field began. William Anderson's *Local Civil Defense in Natural Disaster: From Office to Organization* (1969) was a pioneering study on this topic. That report characterized as uncertain both the roles and the environment in which local crisis-management offices operated. Anderson argued that this uncertainty stemmed from the lack of a consistent resource base for operations, public indifference to the emergency management function, and confusion over organizational bases of authority and task domains. Anderson concluded that the emergency management function for natural disasters was not well-institutionalized in U.S. communities. Civil defense offices in disaster-prone areas that had developed disaster subcultures were an exception to this pattern.

Other early studies found that local agencies charged with emergency management responsibility showed considerable variation in structure, location, and perceived mission. For example, based on extensive research conducted in the 1960s and early 1970s, Dynes and Quarantelli (1977a) concluded that emergency management agencies around the U.S. emphasized different aspects of and approaches to preparedness. Nine different models were identified: maintenance, military, disaster expert, administrative staff, derived political power, interpersonal broker, abstract planner, community educator, and simulation.

In that same study, preparedness activities were found to be fragmented rather than integrated across different organizations and sectors. As a result, organizations tended to plan for disasters in isolation from one another. On the positive side, the scope of preparedness—that is, the different disaster agents that were considered—was broader than it had been previously. The report also noted that over time more community organizations were becoming more interested in planning for disasters, and planning was becoming more integrated. Four general sets of factors were identified that enhanced the legitimacy of local emergency management agencies. The first two were the existence of persistent hazards and the integration of the emergency management office into the day-to-day activities and structure of local government. The other factors judged to be important were the ability of the emergency management office to forge relationships with a range of other community organizations and concrete outputs that emergency management organizations could provide to the community, such as the maintenance of an emergency operations center.

What has been learned in the past 25 years about the quality and effectiveness of the preparedness activities in which local emergency management agencies engage and about the factors that influence preparedness? A follow-up to the Dynes and Quarantelli study described above, which was conducted during the early 1980s, found that local emergency management agencies remain diverse in their organization and operations (Wenger, Quarantelli, and Dynes, 1986). Those agencies vary in a number of ways, including domains and responsibilities, relationships with other emergency-relevant organizations, and resources available to manage disasters. The report judged this diversity to be both natural and desirable, indicating that emergency management agencies are well-adapted to local conditions.

Other studies have also found considerable variability and diversity among local emergency management agencies. The International City Management Association's 1982 survey of more than 6,000 local and county governmental units found considerable structural variation and lack of standardization. For example, communities differed in terms of where in government the emergency management function was located (e.g., in the city manager's office, the fire department, or a specialized unit); in whether the office was independent or embedded in a larger organizational structure; in emergency management staff size; and in whether the emergency management director's position was a full- or a part-time job (Kreps, 1991). Caplow, Bahr, and Chadwick (1984), whose research focused on community readiness for integrated emergency management, made similar points.

With respect to the quality of organizational preparedness efforts, studies suggest that preparedness among local emergency management agencies has improved significantly in U.S. communities since the time of the first assessment. For example, Wenger, Quarantelli, and Dynes, have noted that (1986: 8–9):

> local emergency management agencies appear to be doing a better job at not only producing planning documents (most communities have some sort of a plan) but also at making planning more of a process, rather than a product . . . Most communities have some sort of Emergency Operations Center (EOC), though the quality and adequacy of the facilities varies dramatically . . . Some communities are doing a better job of integrating their disaster planning with those of other organizations . . . Furthermore . . . local communities tend to plan for a wider variety of hazards.

This idea was echoed by Drabek (1993: 5), who concluded that "the extent of disaster planning has increased sharply within both the public and private sectors" and that the quality of that planning has improved as well.

However, while there is general agreement that preparedness capability has increased, we still understand little about the factors that foster successful and effective emergency management units. Drabek (1993: 6) places a good deal of emphasis on the personal attributes of emergency managers themselves, arguing that "[t]he single most significant societal change that has most altered community preparedness has been the increased professionalization of local emergency managers." In his in-depth study of 12 successful emergency managers, Drabek (1990) identified 15 different strategies those individuals employed to keep their agencies on track and to deal with environmental uncertainty, including working to increase constituency support, coalitions, mergers, and joint ventures. What seems most noteworthy about these managerial strategies is the extent to which their focus is interorganizationally- and community-based. Evidently, the most skillful managers are entrepreneurial, outward-looking, and attuned to changes in the multi-organizational and resource environment. Other research, which will be discussed in more detail below in the section on interorganizational and community preparedness, emphasizes the importance of factors such as disaster experience and the properties of community preparedness networks for understanding variations in the emphasis placed on preparedness.

We still know very little about why different emergency management agencies vary in their approaches to emergency preparedness—why, for example, they choose to emphasize particular planning strategies over others. Part of the answer undoubtedly lies in the historical and environmental contexts that have shaped preparedness activities at the community level. Other factors may include emergency managers' judgments about the feasibility and workability of different kinds of strategies and their likely payoffs in the local context.

Along these lines, Kartez and Kelley (1988) obtained data on the views held by 42 managers in Washington State cities and counties toward three planning strategies and their associated activities: managing citizen volunteers, providing disaster information to the media, and working on interorganizational coordination. Those strategies considered most likely to be adopted were those that were seen as having a clear benefit and as involving relatively little effort to implement. For example, volunteer-related activities such as providing training and education to community residents were viewed by respondents as having some potential benefits but also as requiring great effort; this kind of strategy was generally judged to have little chance of actually being carried out. One conclusion the researchers reached (1988: 135) is that "managers judge the prospects of adoption along the path of least resistance."

Despite the fact that preparedness has evidently improved significantly over the past two decades in most U.S. communities, it appears that the disaster planning principles originally identified by Dynes, Quarantelli, and Kreps (1981; see also Kreps, 1991) as crucial for successful planning efforts are still routinely violated by local emergency planners. There remains a tendency to base plans on disaster myths rather than on accurate knowledge, to plan in isolation, and to emphasize command and control. Emergency planners still tend to focus on the written product (i.e., the disaster plan) rather than the planning process, succumb to overconfidence based on successful response to routine emergencies, and use the low priority that others assign to disaster planning as a reason for inaction. Some of these deficiencies can be traced to managers' lack of knowledge of the planning process while others are reinforced by the structure of intergovernmental relationships. For example, the emphasis on the product rather than the more important planning *process*—activities such as holding frequent meetings to assess hazards, developing interorganizational networks, and conducting emergency exercises—is attributable at least in part to the fact that state and federal agency reviewers tend to examine written disaster plans rather than engage in more thorough audits of the preparedness process.

Other Crisis-Relevant Organizations:
Fire, Police, and Emergency Medical Service Providers

Other organizations with significant emergency-related responsibilities, such as fire and police departments, were an early focus of study in the disaster research field (for examples of this work see Drabek and Haas, 1969; Warheit and Waxman, 1973; Wenger, 1973). Later work broadened to consider preparedness and response activities of other key emergency service organizations such as emergency medical service providers and hospitals (Quarantelli, 1983; Tierney, 1985a; Auf der Heide, 1989).

As the focus moves from the emergency management organization to other crisis-relevant organizations, the volume of research on preparedness conducted over the past 25 years drops sharply. For example, aside from research conducted by the Disaster Research Center and a few studies of police department operations in episodes of civil unrest, Wenger, Quarantelli, and Dynes (1989) were able to identify only two other studies that focused specifically on the police in disaster situations.

Consequently, almost nothing new has been learned about police and fire department disaster preparedness. The Disaster Research Center's analysis of police and fire operations (Wenger, Quarantelli, and Dynes, 1989), which was based on extensive field work in communities around the U.S., focused almost exclusively on police and fire response activities in actual disaster situations rather than on pre-disaster planning. The trend in recent research has been toward considering these organizations as elements in larger interorganizational networks—albeit important elements—rather than focusing on their own organizational practices. As a result, we have only the most general idea of how police and fire departments plan for disasters and toward what types of tasks and events this planning is directed.

Although their main interest was in response activities rather than preparedness, Wenger, Quarantelli, and Dynes (1989) did reach some general conclusions about police and fire department preparedness. Police departments, they concluded, tend to devote few resources to emergency planning, although they may be assigned responsibilities in community-wide disaster plans. Larger departments are more likely to plan than smaller ones. When they do plan, police agencies tend to plan internally, in isolation from other community organizations; few have adopted an interorganizational approach to the disaster problem. The police appear to believe that disasters can be handled through the expansion of everyday emergency procedures—that is, they do not consider the qualitative (as opposed to the quantitative) difference between disasters and "everyday" emergencies. Fire departments have improved their preparedness levels and expanded their disaster- and crisis-related tasks beyond fire-fighting. In particular, they tend to be involved in planning for the provision of emergency medical services and for responding to hazardous materials emergencies. Nevertheless, like police departments fire departments show a tendency to plan internally. They:

> . . . continue to be rather autonomous groups that are concerned with maintaining their domain and boundaries . . . Their interaction with other organizations tends to be limited in their daily activities, and this isolation carries over to their planning for other than everyday emergencies (Wenger, Quarantelli, and Dynes, 1989: 115).

Most of what has been learned in the last 25 years concerning the preparedness activities of emergency medical service (EMS) providers— e.g., ambulance companies, paramedic units, and hospitals—comes from studies that were conducted during the mid- to late-1970s, and little new research on EMS preparedness has been done since that time. This research, much of which is now nearly 25 years old, suggests that like police and fire departments EMS organizations tend to plan in relative isolation from broader community-wide preparedness efforts and to see disaster response as primarily an extension of their everyday lifesaving and emergency activities (Quarantelli, 1983). Lack of cohesive EMS planning stems in part from the same kinds of problems that plague the "everyday" provision of EMS: conflicts among the various professions involved in EMS provision, between high- and low-status hospitals, and between public- and private-sector service providers (Tierney, 1985a).

Scattered studies paint a mixed picture of how EMS preparedness efforts have proceeded since those earlier disaster research center studies (Quarantelli, 1983; Tierney, 1985a). For example, looking specifically at EMS preparedness for emergencies involving hazardous chemicals in the state of New York, Landesman (1989) cited research suggesting that hospitals were not prepared to treat victims of chemical disasters, that interorganizational planning efforts were not being undertaken, and that health-care organizations lacked information about which chemical hazards were present in their communities. In contrast, focusing on potential earthquake-related problems, Whitney, Dickerson, and Lindell

(1999) found moderately high levels of earthquake hazard mitigation and preparedness among hospitals they studied in the Southern California region. That research found that public and private non-profit hospitals were better-prepared than ones that operated for profit, and that adoption of earthquake safety measures was correlated with perceived support by senior hospital administrators for seismic risk reduction, as well as with the frequency of hospital disaster coordinators' interactions with their peers at other hospitals.

Auf der Heide's *Disaster Response: Principles of Preparation and Coordination* (1989) contains both syntheses of research on EMS preparedness and planning guidelines. Although the book is intended to serve mainly as a guide for practitioners, it also contains discussions of a number of factors that research indicates have contributed to low levels of EMS preparedness. Among these factors are the lack of public awareness, the tendency to underestimate disaster probabilities, overreliance on technological fixes, and fatalism and defeatism. Other negative influences include the lack of governmental support for preparedness, lack of an organized constituency supporting preparedness, competing priorities, and the difficulty of substantiating the benefits that derive from preparedness. Finally, Auf der Heide argues that low levels of preparedness can be attributed to inflated expectations about response capability, ambiguities about responsibility for preparedness, and the prevalence of what he terms the "paper plan syndrome."

Some other recent publications, while not based directly on healthcare preparedness research, are also relevant to this topic. Noji's edited volume, *The Public Health Impacts of Disasters* (1997), summarizes findings on the ways in which various types of disasters affect mortality and morbidity worldwide. Landesman's book, *Emergency Preparedness in Health Care Organizations* (1996), provides guidance and case study material on hospital emergency planning and on the provision of emergency medical services during disasters.

When those early studies were conducted the fiscal crisis that is currently creating so much difficulty for health care providers still was years away. Perhaps more so than most other crisis-relevant organizations, those in the EMS sector—particularly hospitals—face an uncertain and generally unsupportive environment. No research has been conducted on how the crisis in health care is affecting planning for disasters, but since disasters (as opposed to "everyday" emergencies and possibly mass-casualty incidents) never were a major priority for most EMS organizations we can only assume that they have moved downward on the agenda, rather than up.

Government Organizations Without Disaster-Relevant Missions

Comparatively little research attention has been devoted to emergency preparedness among government organizations that do not perform crisis-relevant functions (Lindell and Meier, 1994). Such preparedness is nevertheless important for assuring that governments will be able to continue to operate following disasters, when departments and agencies are called upon to respond not just to disaster-generated demands but also to continue to meet "normal" demands (Wolensky and Wolensky, 1991; Anthony, 1994; Cooke, 1995). Virtually all existing research on such agencies has been conducted on municipal and county organizations.

Three factors have been consistently identified as positive correlates of organizational preparedness among non-emergency organizations. First, organizational size has been identified as positively related to emergency planning activity (Quarantelli, 1984). Larger organizations have more resources and are also likely to have a greater perceived need for strategic planning, and this need is correlated with a concern for emergency preparedness (Gillespie and Streeter, 1987; Banerjee and Gillespie, 1994). Second, the level of perceived risk among organizational and department managers is positively correlated with emergency preparedness (Mileti, 1983; Mileti and Sorenson, 1987; Drabek, 1990). Finally, the extent to which managers report seeking information about environmental hazards is positively correlated with organizational preparedness (Lindell et al., 1996a, 1996b; Barlow, 1993). Perry and Lindell (1997) assembled these factors into a model predicting earthquake preparedness by municipal and county departments. The three variables ultimately explained about two-thirds of the variance in earthquake preparedness, with risk perception and self-reported information-seeking behavior being the most important.

Studies of Private Sector Organizations

When a disaster strikes and businesses are damaged or either temporarily or permanently unable to continue operating, both those businesses and the local economy suffer. Employees may face losing their livelihoods, and community residents may find themselves forced to search elsewhere for goods and services that were easily available before the disaster. When central business districts suffer concentrated disaster damage—as occurred, for example, in the 1974 Xenia, Ohio, tornado, the

1983 earthquake in Coalinga, California, and the 1989 earthquake in Santa Cruz, California—communities face a host of problems, including the potential for permanent loss of businesses, loss of sales and property tax revenues, and the need to finance commercial recovery. Thus, it is important to know more about the extent to which businesses are prepared for disasters and which types of businesses are most likely to prepare and why, as well as how to encourage businesses to undertake protective measures.

Unfortunately, few studies in the literature have addressed emergency preparedness activities in the private sector. In fact, until relatively recently, business disaster preparedness had been virtually ignored by researchers. When businesses were studied there was a tendency to focus on specific kinds of private-sector organizations, such as tourist-oriented firms (Drabek, 1991a, 1994, 1995) and hazardous materials producers and handlers (Quarantelli et al., 1979; Gabor, 1981; Lindell and Perry, 1998). Additionally, many studies on business preparedness employed small samples (Barlow, 1993) or concentrated on preparedness for specific types of disaster events (Mileti et al., 1993), which has also limited the extent to which their findings can be generalized. Studies on the preparedness activities of large and representative groups of businesses only began to appear during the latter half of the 1990s (Dahlhamer and D'Souza, 1997; Webb, Tierney, and Dahlhamer, 2000).

The research that does exist indicates that private firms are less than enthusiastic about emergency preparedness. For example, Drabek (1994) assessed the quantity and quality of evacuation preparedness among the 185 tourist-oriented firms he studied as "unsatisfactory" overall; preparations were judged adequate for only 31 percent of the businesses in his survey. Mileti and his associates (1993) found that fewer than half of the businesses they interviewed in the San Francisco Bay Area, which has both a high probability of earthquakes and recent earthquake experience, had engaged in recommended preparedness activities such as developing disaster plans, training employees, and conducting drills. A recent study of business preparedness in Memphis and Des Moines that employed a preparedness checklist similar to those used in research on households also found generally low levels of preparedness. In Memphis, for example, businesses had on the average engaged in four out of 17 recommended preparedness measures; in Des Moines, the average was 1.7 out of 13 possible activities (Dahlhamer and D'Souza, 1997).

With respect to factors that influence preparedness among private-sector organizations, certain structural characteristics of firms appear to

This North Carolina bank suffered major damage in a 1998 tornado. Researchers are increasingly turning their attention to the ways in which disasters affect private-sector organizations and to private-sector disaster preparedness.

be the strongest predictors of preparedness levels. The characteristic most consistently related to preparedness is firm size. For example, Quarantelli et al. (1979), in a study of chemical companies in 18 communities around the U.S., found that larger companies had engaged in more extensive planning than their smaller counterparts. Drabek (1991a; 1994) also found size to be positively related to the extensiveness of evacuation planning among two samples of tourist-oriented firms, and in their study of businesses in Memphis and Des Moines, Dahlhamer and D'Souza (1997) found the same positive relationship between size and preparedness levels. In a study conducted after the 1994 Northridge earthquake, Lindell and Perry (1998) also found that size was positively associated with the adoption of earthquake mitigation and preparedness measures among Los Angeles hazardous materials-handling firms.

The age of a business is a second characteristic that appears to be related to a firm's preparedness, but there is some question about whether and how age makes a difference. While Drabek (1991a) found that firms that were in business for six or more years were more likely than younger firms to emphasize preparedness, the Quarantelli study on chemical emergency preparedness (Quarantelli et al., 1979) found that preparedness was higher among newer companies, and Dahlhamer and D'Souza (1997) found that the length of time a firm had been in business had no significant impact on preparedness in either of their samples.

There is also some evidence suggesting that nationally-based companies with multiple locations have higher levels of preparedness than individual local firms (Drabek 1991a, 1994, 1995), perhaps because parent companies can both mandate preparedness for their units and supply relevant resources. However, this finding is contradicted by the Dahlhamer and D'Souza study (1997), which found that whether a business was an independent proprietorship or part of a chain had no effect on preparedness levels.

Preparedness may also vary according to business type, but here again the evidence is not consistent. In his research on the tourist industry, Drabek (1991a, 1995) found that lodging establishments were more likely to engage in evacuation planning than restaurant, entertainment, and travel firms. In their analysis of earthquake preparedness among 54 firms in the San Francisco Bay Area, Mileti and his associates (1993) found preparedness to be higher among health, safety, and welfare organizations that had staff specifically devoted to earthquake planning. Dahlhamer and D'Souza (1997) reported that firms in one particular sector—finance, insurance, and real estate—had higher levels of preparedness in Memphis, but that sector was not a factor in Des Moines, the other city they studied.

As noted above, there is evidence suggesting that disaster experience contributes to higher levels of household and community preparedness. The same pattern may well hold for businesses. Prior experience exerted a positive influence on pre-event preparedness among businesses in Memphis, Des Moines, and Los Angeles (Dahlhamer and Reshaur, 1996; Dahlhamer and D'Souza, 1997). Preparedness has also increased among businesses in areas that were hard hit by the Loma Prieta earthquake in 1989 and by Hurricane Andrew in 1992 (Webb, Tierney, and Dahlhamer, 2000).

While disaster research has focused on the preparedness activities of organizations since the field began, until very recently that work has centered almost exclusively on public sector organizations rather than on the private sector. Indeed, one of the most noteworthy trends in recent research has been the increasing focus on such topics as business preparedness, the ways in which disasters can affect businesses, and business recovery following disasters. This research area is still in its infancy, however, and as the discussions above demonstrate, the studies that have been conducted to date have yielded few conclusive findings.

INTERORGANIZATIONAL AND COMMUNITY
DISASTER PREPAREDNESS

Numerous studies have shown that local support for disaster preparedness is low in most communities, that emergency planners tend to
have low prestige, and that relatively few resources are allocated to disaster preparedness and response (Labadie, 1984; Rossi, Wright, and
Weber-Burdin, 1982). Disasters do not receive a higher priority because
such events are infrequent in any given locality, responders tend to
overgeneralize from their experiences with routine emergencies, and non-
specialists tend either to underestimate the magnitude of disaster demands (resulting in unrealistic optimism) or grossly overestimate them
(resulting in fatalism). Moreover, as Gillespie (1991) has noted, there are
a number of organizational obstacles to the development of coordinated
systems of community emergency preparedness. These include the tendency of organizations to seek autonomy, staff commitment to professional ideologies, differences in organizational technologies and resource
needs, fears about the loss of organizational identity, concerns about loss
of scarce resources and the proliferation of organizations and interest
groups across political jurisdictions, and perceived differences in the costs
and benefits of cooperation.

We noted earlier that one of the key research advances in the past
25 years has been the development of systematic approaches to collecting
data on preparedness at the household level. Similar progress has been
made in the conceptualization and measurement of interorganizational and
community preparedness. Drabek's (1987, 1990) research suggests that
community differences in emergency preparedness can be attributed at least
in part to the effects of the strategies and structures adopted by local emergency managers as well as to differences in the interorganizational structures of their disaster planning networks. However, methodologically sophisticated studies of emergency preparedness networks remain very rare
in the literature. And, like research on households, community studies
show considerable variation both in their approaches to measurement and
in the variables used. Moreover, as with research in the field generally,
studies on community preparedness have tended to focus either on particular types of disaster agents or on particular types of preparedness networks, again raising the question of generalizability.

One major contribution to the specification and measurement of the
properties of preparedness at the organizational and interorganizational
levels was stimulated by research on emergency preparedness in St. Louis

(Gillespie and Streeter, 1987; Gillespie et al., 1993). These researchers conceptualized emergency preparedness as involving two basic components, physical preparedness and social preparedness (Gillespie et al., 1993). Physical preparedness centers on taking actions to ensure that structures and facilities can withstand disaster impact and that buildings and their contents do not become a life safety threat when a disaster strikes. The social dimension of preparedness involves taking actions to ensure that community organizations are able to respond to the needs of victims in the event of a disaster. Those actions include understanding what state and federal programs are available at the time of disaster, planning for situations involving warning and evacuation, establishing emergency record-keeping systems, and developing disaster plans and mutual aid agreements. Social preparedness is further defined as having planning, training, financial, and community education or community involvement components. Both physical and social preparedness are seen as part of a five-phase preparedness cycle that consists of raising awareness, conducting hazard and vulnerability assessments, improving knowledge about hazards and how to cope with them, planning, and practice.

Focusing on 80 public and private organizations, Gillespie and Streeter (1987) found that emergency preparedness was positively associated with five variables.* Two were related to the internal structure of organizations. These were organizational capacity, which was defined as the number of different emergency services an organization could provide, and formalization of roles and procedures within the organization. The other important predictors of preparedness were disaster experience, the existence of a disaster subculture, and the quality of interorganizational relations, as measured by the formalization of interorganizational agreements and exchange of services, resources, and personnel.

In a study that is virtually unique in the literature, Gillespie and his colleagues (Gillespie et al., 1992) used sophisticated network-analytic procedures and advanced statistical methods to analyze relations among 160 organizations constituting the emergency preparedness network in St. Louis. They identified four network properties—cohesion, interorganizational contact, autonomy, and density of contacts—that either enhance or discourage preparedness. On the positive side, the greater the

---

* In this study, preparedness activities were operationalized using measures of organizational planning, training, and familiarity with the integrated emergency management system, a model of emergency management that was recommended at the time the study was conducted.

network cohesion and number of interorganizational contacts that occur, the higher the level of preparedness. However, the greater the autonomy of organizations in the network and the greater the density of contacts, the lower the preparedness. The total number of contacts, which was positively related to preparedness, was negatively correlated with both autonomy and density. Interestingly, this research also found that the factors that enhance preparedness are different for different types of organizations. For example, for social service agencies, more contacts with different types of organizations generally led to higher preparedness levels, but this was not the case for emergency management agencies. Equally interesting were the substantive findings on network composition this study yielded. Several clusters were identified, including a central seven-organization cluster, more peripheral organizational groupings, and a cluster consisting primarily of crisis-relevant organizations such as police and ambulance providers.

This research suggests that network structure has an important independent influence on how constituent organizations approach preparedness. In other words, while researchers must carefully study the activities and attributes of individual organizations and their leaders—their position within local government, what resources they have, the priority their management places on preparedness, and so on—at the same time, it is also necessary to take into account the larger structures in which these organizations are embedded and how their network positions affect their operations. (For other discussions on the measurement and analysis of community preparedness see Gillespie and Streeter, 1987; Gillespie, 1991; Gillespie et al., 1992, 1993.)

Network analysis appears to be a particularly promising approach to the study of interorganizational and community preparedness. Its use could help link the study of preparedness to the broader literature on the properties and impacts of interorganizational networks, including their effects on the perceived power of constituent organizations and on organizational effectiveness (see, for example, Knoke, 1990), which could in turn provide important insights for the analysis and interpretation of data on preparedness and response. However, little work of this kind is being done in the field. In short, an important set of theoretical, conceptual, and methodological tools for studying emergency preparedness in an interorganizational context is being almost totally ignored.

A considerable amount of community emergency preparedness research has focused on problems associated with the manufacture and handling of hazardous materials. In the late 1970s and early 1980s, the

Disaster Research Center conducted research on interorganizational and community preparedness and response in emergencies involving hazardous chemicals. Part of that project involved assessing preparedness in 19 communities around the U.S. that were selected because of their relatively high vulnerability to chemical accidents. The study found that key emergency response organizations in the communities studied were generally aware of the problems posed by chemical hazards in their communities. However, considerable local variation was observed in the interorganizational arrangements that had developed to prepare for those kinds of emergencies. At one end of the continuum were communities in which virtually no serious planning for such emergencies was taking place, and almost nothing was being done to mitigate the dangers posed by chemical hazards beyond the minimum required by regulations. At the other were communities in which reasonably well-integrated mutual aid systems that included both industry and governmental emergency responders had developed and become institutionalized. The largest group of communities fell in the middle: planning for chemical emergencies was undertaken by a loosely-structured, primarily informal network of organizations spanning both industry and government.

The study identified a number of significant weaknesses in the planning that was being undertaken for chemical emergencies. In many of the communities studied, preparedness was found to be (Quarantelli, 1981a: 83) "nonexistent, poorly developed, or merely a paper plan." Even in the communities judged to be best-prepared, important concerns such as evacuation planning for community residents should a chemical release occur were not being addressed. At the facility level planning was quite uneven, and the facilities that probably most needed to plan—older plants located near population centers—appeared to be the least likely to do so. On the other hand, the newer facilities and those located in industrial parks where they presumably posed less of a hazard to the population were more likely to plan. Compared with planning for natural disasters, chemical emergency planning had to overcome additional barriers to communication, coordination, and cooperation between private plant operators and the larger community preparedness and response network. At the same time, the expertise and many of the resources required for responding to such emergencies were generally concentrated within the industry itself rather than in the larger community. Among the study's (Quarantelli, 1981a: 75) conclusions were that:

> Not only does planning for chemical disasters suffer from the problems attendant to all general disaster planning in American communities, but

it also has additional problems of its own. In particular [it] is plagued by the public-private sector division in our society, and also by the fact that the local community (which necessarily has to be the first responder) has generally less capability and knowledge for dealing with chemical emergencies than do extra- and super-community social entities.

Advances in research on community preparedness for hazardous materials emergencies have been stimulated by the passage of the Emergency Planning and Community Right to Know Act, Title III of the Superfund Amendment and Reauthorization Act (SARA) of 1986. SARA Title III, which was signed into law in part as a consequence of the 1984 Bhopal disaster, was intended to address problems like those identified in the earlier DRC study on chemical emergency preparedness. The law included provisions that required facilities to disclose information on the hazardous materials they were manufacturing, processing, and storing, and it also mandated the creation of local emergency planning committees (LEPCs) nationwide to (among other things) prepare for hazardous chemical releases.

Evidence suggests that at the time the law was enacted communities were not well-prepared for major emergencies involving hazardous chemicals. Research by Sorensen and Rogers (1988), conducted just after the LEPC program began, focused on one aspect of preparedness—the ability of communities to warn the public should an incident occur involving a hazardous chemical facility. The study sample consisted of local emergency planning agencies that were selected because facilities identified as hazardous by the U.S. Environmental Protection Agency were located in their communities. Among the study's findings were that communications links between facilities and communities were not reliable; that community agencies were unclear about how warnings would be communicated to them by facilities; that about half the communities had procedures that were unclear about what would be done once a warning was received from a hazardous facility; and that communities were unclear about what information they would need if a chemical accident were to occur. Based on their survey the authors (Sorensen and Rogers, 1988: 104) concluded that:

> ... few communities had well-developed plans and procedures to guide emergency response. Notably lacking were capabilities to make decisions. Both lack of procedures and, more basically, insufficient knowledge about what information is needed to make a decision suggest major problems with issuing a timely warning.

A number of studies on LEPCs followed, but there still is not a great deal known about their implementation and effectiveness. On the one hand, Feldman (1993) cites research indicating that by 1990 86 percent of LEPCs were functioning nationally, and just under half had developed and submitted the required response plans. On the other hand, looking at approximately the same time frame, Solyst and St. Amand (1991) found that only 19 of 56 states and territories subject to SARA Title III had developed emergency response plans. Lindell and Meier (1994), focusing on LEPCs in Michigan, found that there was considerable variation in the extent to which LEPCs had completed key preparedness tasks such as conducting hazard analyses (about 30 percent of the LEPCs reported having done so), developing site-specific emergency plans (26 percent), and training emergency response personnel (15 percent). Although not much progress appeared to have been made at the time of their study, these authors argue that LEPC preparedness activities should be judged in the context of mitigation and preparedness planning for other hazards, which is also often found lacking.

Lindell and his colleagues have also found that even though greater community resources, availability of funding, and higher levels of hazard vulnerability are associated with higher levels of LEPC preparedness, none of the correlations is large. These modest correlations are consistent with the inconclusive effects of contextual variables found by other recent investigators (Adams, Burns, and Handwerk, 1994). Instead, the ability of LEPCs to plan effectively appears to be influenced more by specific organizational characteristics of the LEPC, such as staffing and structure and emergency planning resources, than by contextual characteristics such as community hazard vulnerability and resources (Lindell, 1994a; Lindell and Meier, 1994; Lindell and Whitney, 1995; Lindell, Whitney, Futch, and Clause, 1996a, 1996b).

PREPAREDNESS AT THE STATE AND NATIONAL LEVELS

State government has been described (Durham and Suiter, 1991: 101) as "the pivot in the intergovernmental system . . . in a position to determine the emergency management needs and capabilities of its political subdivisions and to channel state and federal resources to local government." States possess broad authorities and play a key role in emergency preparedness and response, both supporting local jurisdictions and coordinating with the federal government on a wide range of disaster-related tasks. Federal resources cannot be mobilized in a disaster situation without a formal request from the governor, and states have a number of their own resources at their disposal for use in emergencies, including the personnel and resources of the National Guard. States are required to develop their own disaster plans, and they typically also play a role in training local emergency responders. States have significant responsibilities for environmental protection and the delivery of emergency medical services, and state emergency management duties have broadened in recent years as a result of legislation like SARA Title III, which requires states to coordinate the chemical emergency preparedness activities of LEPCs.

In view of the important roles states play in the management of hazards and disasters, the vanishingly small amount of research focusing on state-level emergency preparedness activities is surprising. In the mid-1980s, Drabek's *Human System Responses to Disaster* (1986), a compendium of research findings in the disaster area, made no mention of state-level planning and contained virtually no material on the state as a separate unit of analysis in disaster research. A decade later, Waugh and Sylves (1996: 49) noted that "[t]he structure and operation of state-level emergency management has gotten far too little attention, despite the criticism of performance in recent disasters."

The first effort to focus on comprehensive disaster management from the state perspective was undertaken in the mid-1970s by the National Governors' Association (NGA). One key focus of the NGA project was on compiling information on federal policies, legislation, and disaster assistance programs for use by governors (National Governors' Association, 1978a, 1978b, 1979). However, the study also obtained detailed information from 57 states, commonwealths, and territories on their emergency preparedness and response activities.

The NGA study found that, like their local counterparts, state-level emergency management agencies were located within various offices and branches of state government. The five most commonly-identified organizational locations for the disaster management function were (in descending order of frequency): as a bureau under an adjutant general; as a division under a civilian department; within the office of the governor; under the state police; and under a state council. The report suggested that, while organizational location is probably a factor in the effectiveness of a state-level emergency management agency, its relationship with the governor's office is likely to be even more important. Where governors are concerned about and supportive of disaster management activities, state agencies have a better chance of being effective (National Governors' Association, 1978c). Although NGA reports do not contain detailed information on state-level preparedness activities, they do suggest that at the time the study was conducted states varied in both the quantity and quality of their preparedness; that the organizational location of the emergency management office was a factor in preparedness; and that states were generally better-prepared for natural than for technological disasters.

Since that time, there have been only a few scattered studies of limited scope on preparedness measures undertaken by states. For example, one series of descriptive reports contains sketchy information on the disaster-related responsibilities undertaken by state emergency medical services authorities (National EMS Clearinghouse, 1988). A few other studies (Rossi, Wright, and Weber-Burdin, 1982; Drabek, Mushkatel, and Kilijanek, 1983; Mittler, 1989) report findings on state officials' attitudes regarding the importance of the disaster problem and what should be done about it and also on the hazard-reduction actions taken by states. However, those studies focus primarily on mitigation rather than on preparedness or response. Similarly, Olshansky's (1994) work on states and earthquake hazard reduction focused primarily on mitigation.

Surveys like the one discussed above on LEPCs (Solyst and St. Amand, 1991) examine state-level preparedness initiatives involving a particular type of disaster agent, rather than looking at emergency preparedness generally. Along the same lines, May and Williams (1986) assessed preparedness planning for earthquakes in 22 seismically-vulnerable states as part of a larger study on disaster policy implementation. At the time their study was conducted most states were aware of the earthquake hazard and had some preliminary vulnerability data. Only a small number of states had begun incorporating earthquakes into their disaster plans, and one-third of the sample had attempted to establish earthquake task forces at the state level. May and Williams described state-level earthquake preparedness efforts outside California as "sporadic at best" (1986: 99).

In the more than 20 years since the NGA report, no studies have been conducted to reassess state disaster preparedness activities in a comprehensive fashion. Although state-level preparedness typically is touched upon in studies undertaken following individual disaster events, it has not been the focus of systematic, comparative study. As a consequence, little can be said with confidence about where states currently stand with respect to emergency preparedness or about the factors that influence preparedness at the state level. Researchers concerned with the operation of the intergovernmental system (Mushkatel and Weschler, 1985; Waugh and Sylves, 1996) point to various topics on which research is needed. These include the impact of structural arrangements—such as the organizational location of the emergency management—on how states plan for and respond to disasters and the extent to which agreements between the federal government and the states have fostered improvements in emergency management. It is clearly incorrect to view states merely as a "pass through" for federal assistance in disasters or as unimportant players in the management of hazards. States can take an active or a passive role in promoting preparedness and response, and what they do undoubtedly makes a difference at the local level. However, without research that takes an in-depth look at what states are actually doing, researchers can conclude little about their role in the preparedness process.

The picture is scarcely better at the national level of analysis. Drabek has called attention to what he terms a "void in the empirical data base" (1986: 61) concerning disaster preparedness at the national level. Much of the knowledge that we have of federal government preparedness comes from detailed case studies that either focus on the federal government at a particular point in time or that assess changes in federal policies and programs that have taken place over time. For example, May and Williams's (1986) excellent study on disaster policy implementation in the intergovernmental system contains case study material on two preparedness- and response-related programs, earthquake preparedness planning and the ill-fated crisis relocation planning program (the other two programs discussed are flood plain management and dam safety regulation). Another important piece of research is Kreps's (1990) history of the evolution of federal emergency management policy since World War II. Included in that analysis are discussions of the almost continual reorganizations the emergency management system has undergone at the federal level, the trend toward broadening federal programs over time to include more hazards and different forms of assistance, and the persistence of the "dual use" orientation, which emphasizes preparedness for both war-related emergencies and disasters (see also Clary, 1985; Drabek, 1991b; and National Academy of Public Administration, 1993; as well as our discussion in Chapter Six for more information on changes in federal hazard management policies).

One general research-based observation concerning national-level preparedness is that "[i]n all societies, disaster planning will be uneven nonuniform across hazard types, reflecting cultural values, assumptions, and power differentials" (Drabek, 1986: 60). We can see this pattern played out in the U.S. where, reflecting their Cold War origins, federal preparedness initiatives have been shaped historically by national defense considerations. Federal emergency preparedness evolved out of an earlier concern for civil defense, and for many years planning for nuclear war persisted as a key element in federal preparedness activities. At different times, this emphasis has made implementing preparedness measures difficult. For example, an important factor in local resistance to federally-directed "crisis relocation planning" for disasters was that program's relationship to nuclear war readiness (May and Williams, 1986; Waugh, 1988). More recently, federal preparedness initiatives have reflected growing concerns about the domestic risks posed by terrorism and weapons of mass destruction.

Drabek (1986) has also noted that national-level preparedness initiatives are shaped to a considerable extent by dramatic disaster events. For example, the Three Mile Island nuclear accident stimulated federal action to encourage extensive evacuation planning for areas around nuclear power plants (Sylves, 1984). As noted earlier, the Bhopal disaster was a major factor influencing the content of SARA Title III, which mandated the creation of LEPCs. Similarly, many provisions in the Oil Pollution Act of 1990 came about as a direct response to the 1989 *Exxon Valdez* oil spill. The federal response plan already had been developed prior to the 1992 occurrence of Hurricane Andrew, but the delayed and uncoordinated response to that disaster prompted strong criticisms and calls for improved response planning. (The federal response plan is discussed in more detail in Chapter Four.)

One key theme in the research literature is that federal preparedness is influenced and constrained not only by institutional power differentials but also by the nature of the intergovernmental system itself. Based on analyses by researchers and agencies like the U.S. General Accounting Office, Waugh (1988) has pointed to a number of factors that have made the implementation of federal preparedness initiatives difficult. These barriers include the sheer complexity of the intergovernmental system, a lack of leadership at the federal level—due in part to the weak position held by the Federal Emergency Management Agency (FEMA)—and poor federal interagency cooperation. To the extent preparedness actions are based on detailed hazard analyses, they may be difficult to implement from a technical standpoint. The goals and objectives to be pursued at the federal level are not always clearly articulated, the resources are frequently insufficient, and federal preparedness lacks a strong constituency.

Focusing specifically on federally-mandated preparations for emergencies at nuclear power plants, Aron (1990) outlined a host of barriers that stood in the way of effective implementation: jurisdictional complexity that made planning difficult; insufficient funds and technical expertise; the reluctance of many local governments to participate; conflict between FEMA and the Nuclear Regulatory Commission (NRC) about responsibility for directing and assessing preparedness activities; lack of clearly-specified roles among the various agencies charged with emergency duties; and constitutional issues regarding the relative power and authority of federal, state, and local jurisdictions. Nevertheless, Aron (1990: 216) concluded that, "[s]hould a serious accident occur, we are far better prepared than we were a decade ago to provide protective action for affected citizens."

It is not the purpose of this book to compare preparedness and response activities cross-nationally. However, our understanding of U.S. emergency management policies and practices would improve considerably if we had better information on the organization and effectiveness of disaster preparedness and response in other countries. Little systematic research exists comparing the organizational features, policies, and practices of national governments. However, even anecdotal evidence indicates that national emergency management strategies vary considerably, and what little we do know raises a number of questions. For example, the U.S. governmental system is a decentralized federal system, and the organization of emergency management activities reflects that decentralization. In the U.S., we believe that allowing responsibility for managing emergencies to reside at the local level provides the best way of ensuring that emergency management organizations act in ways that are responsive to local needs. The U.S. pattern of organization would seem to be particularly well-suited to situations in which there is sufficient capacity at the local level to handle emergency-related demands. However, what about nations that do not divide powers and authorities among different governmental levels, or countries that do not have sufficient resources at the local level? Other nations, such as Japan, are more highly centralized governmentally. Does centralization enhance or inhibit effectiveness? Perhaps better stated: under what conditions and for what tasks does centralization work best?

Similarly, in some societies the military plays a prominent role in disaster management, while in others it is kept in the background. In Japan, for example, the public frowns on the use of the military in disaster, and distrust of the military was one reason its personnel and equipment were not well-utilized following the 1995 Kobe earthquake. In countries that are currently ruled by authoritarian military regimes or that have been in the past, the military may possess substantial resources yet be an object of public fear and hatred. After Hurricane Andrew some

U.S. observers called for greater disaster involvement by the military in domestic disaster situations, while others criticized the role played by the armed forces. What role can and should the military play in disaster preparedness and response?

During the Cold War we knew very little about Soviet hazard management policies. Since the dissolution of the Soviet Union we know even less about emergency preparedness in the countries that were formerly within the Soviet sphere of influence. What is the status of disaster preparedness in the former Soviet-bloc countries, and what lessons can the experience of these nations teach the U.S.?

Research comparing countries in Africa and Latin America suggests that the political ideologies that governments favor are related to the approaches they take to managing hazards, which are related in turn to casualty rates and economic losses (Seitz and Davis, 1984). Can these patterns be observed in other societies as well? Since no studies have been conducted to replicate or extend this work, we don't know. Both theory and practice would likely benefit from cross-cultural research on national emergency preparedness and overall disaster management policy.

For the purposes of this discussion, it is sufficient to emphasize that emergency preparedness at both the state and national levels has been seriously understudied, receiving only cursory attention in the literature. In Chapter Six we will return to a discussion of the ways in which the roles national and state governments play in the U.S. intergovernmental system influence disaster preparedness and response.

RESEARCH ISSUES AND
UNANSWERED QUESTIONS

Twenty-five years ago, what we knew about preparedness was based overwhelmingly on studies of emergency-relevant organizations, such as offices of civil defense and fire and police departments. Major advances have been made in the study of household preparedness, and we now have much more detailed information both on core disaster agencies and on other types of organizations than we had a generation ago. More recently, researchers have begun to focus on the private sector, which was totally neglected in earlier studies. The involvement of researchers with an interest in interorganizational relations and the development of new types of preparedness networks, such as LEPCs, have helped to stimulate an important new interorganizational focus in the study of preparedness. Despite these major advances, old questions remain unanswered and new ones have emerged. We next discuss several lingering issues that have yet to be adequately addressed and that suggest several avenues for further research.

Does Preparedness Make A Difference? If So, How and Why?

The question of whether high levels of preparedness do in fact improve the ability to respond effectively during actual disaster situations is central to the study of disaster preparedness and response. Reviews of earlier research in the field (Mileti, Drabek, and Haas, 1975) concluded that this was indeed the case. From a more practical point of view, preparedness should be easier to promote if its effectiveness can be demonstrated empirically. For these reasons, one would expect research on the link between preparedness and response to be very prominent in the literature. However, while the topic has not been neglected totally, neither has it received a high degree of research emphasis. One reason may be that the issue is a relatively difficult one to address, since it requires not only analyzing and assessing responses to specific disaster events but also having data on the nature and extent of preparedness prior to those events (Banerjee and Gillespie, 1994).

At the household level, pre-planning does seem to foster adaptive behavior. Indeed, with respect to household units, the argument has been made that it is "almost axiomatic that higher levels of preparedness will result in more appropriate response activities" (Banerjee and Gillespie, 1994: 345). This has been shown to be particularly true for evacuation in the face of an impending threat (Perry, 1979a; Perry et al., 1981; Perry and Greene, 1982, 1983).

Less is known about the impact preparedness has on response effectiveness at the organizational and community levels. On the one hand, a number of studies suggest that it does make a difference; other things being equal, organizations that have engaged in prior planning perform better in actual emergencies than those that have not (Mileti, Drabek, and Haas, 1975; Saarinen and Sell, 1985; Kartez and Lindell, 1987, 1990). On the other, there may be a tendency to deduce incorrectly that pre-planning was adequate from the fact the response to an actual event worked well. Actually, it may be as Kartez and Kelley (1988: 129) suggest: "[t]he fact that local managers and agencies have adapted in the event is not evidence of preparedness, only of ingenuity and fortune."

Preparedness and response effectiveness may vary independently of one another. A Disaster Research Center study on the preparedness and response activities of local emergency management agencies conducted in the early 1980s found that, while preparedness for disasters had improved markedly in U.S. communities, the response to actual disaster events had not (Wenger, Quarantelli, and Dynes, 1986). Quarantelli later observed that among the conclusions that can be drawn from this study is that "even if preparedness is good, it does not follow that managing a disaster will also be good . . . Good planning does not automatically translate into good managing" (Quarantelli, 1993: 33).

This is also the conclusion that was reached more recently by Clarke (1999: 57), who found that, despite what we would like to believe, in many crisis situations "planning and success *do not* coincide but are loosely connected or even decoupled entirely" (emphasis in the original). Not only does planning sometimes prove ineffective, Clarke notes, but we can also point to disasters that were well-handled in spite of an apparent absence of planning or failure on the part of organizations to employ existing plans.

In commenting on disasters that have been managed well, researchers often observe that the disaster-stricken community in question had engaged in extensive emergency planning prior to the event. Yet we less frequently look in detail at whether response effectiveness was the result of effective planning, emergency period improvisation, or sheer good luck. How much variation in response effectiveness can be attributed to pre-event planning and how much is due to other factors, such as the length of the warning period, the quantity of resources that happen to be on hand, or even the time of day? Following the 1994 Northridge earthquake it was widely acknowledged that, had the temblor not occurred at around 4:30 in the morning on a national holiday, responding organizations would have faced many more severe challenges than that event presented. Was the fact that the city of Los Angeles responded so effectively to the earthquake the result of the planning that had been done beforehand—which was extensive—or due to the fact that the problems that developed did not really tax the response system? These kinds of questions are difficult to address, in part because disaster research does not lend itself well to comparative research in which important factors such as disaster agent characteristics, severity of impact, community characteristics, and the nature and comprehensiveness of planning efforts can be systematically varied.

Can We Identify Optimal
Preparedness Strategies?

Keeping these cautions in mind, it still seems reasonable to assume that planning should have a positive effect on organizational performance in crisis situations. But that also leads logically to questions about what constitutes good planning and what aspects of planning are most likely to make a difference. From both a theoretical and a practical standpoint it is important to determine whether particular organizational strategies or approaches foster more comprehensive preparedness efforts. Is there, in other words, an optimal way to organize for disaster response? Drabek's (1987, 1990) finding that some successful emergency managers enthusiastically endorsed strategies that were explicitly rejected by other equally successful managers suggests there is no single best way to organize for emergency preparedness.

Nonetheless, there does seem to be considerable support in the literature for the idea that some planning models and approaches are better than others, largely because they do a better job of preparing organizations for meeting the demands posed by a disaster. One major theme in the disaster literature is that response-related problems have their origins in part in planning that makes incorrect assumptions about how disasters should be managed. Dynes (1993, 1994) contrasts two ideal-types of planning and response frameworks, termed the military and the problem-solving models, respectively (see Table 2.2). The former sees disasters as chaotic situations in which social disorganization is so widespread that centralized, command-and-control-oriented strategies must be implemented. In contrast, the latter approach assumes that disaster-stricken communities possess sufficient resources and problem-solving capacity to cope without the imposition of hierarchical authority and that the goal of preparedness and response efforts is to help develop and use those capabilities.

Synthesizing recommendations from a long tradition of work on disasters, Quarantelli (1982d) has identified ten general principles of disaster planning that are applicable to a range of planning efforts, whether carried out by governments, private-sector organizations, or other social units. These are that planning: (1) is a continuous process; (2) entails attempting to reduce the unknowns in the anticipated disaster situation, although it is impossible to pre-plan every aspect of a response; (3) aims at evoking appropriate (not necessarily rapid) response actions; (4) should be based on what is likely to happen and what people are likely to

TABLE 2.2 Contrasting Models for Emergency Preparedness and Response

Military Model	Problem-Solving Model
1. It assumes social chaos and dramatic disjunctures during the emergency.	1. Emergencies may create some degree of confusion and disorganization at the level of routine organizational patterns, but to describe that as social chaos is incorrect.
2. It assumes the reduced capacity of individuals and social structures to cope.	2. Emergencies do not reduce the capacities of individuals or social structures to cope. They may present new and unexpected problems to solve.
3. It creates artificial social structures to deal with that reduced capacity.	3. Existing social structure is the most effective way to solve those problems. To create an artificial emergency-specific authority structure is neither possible nor effective.
4. It expresses a deep distrust of individuals and structures to make intelligent decisions in emergencies.	4. Planning efforts should be built around the capacity of social units to make rational and informed decisions. These social units need to be seen as resources for problem solving, rather than as creating problems themselves.
5. It places responsibility in a top-down authority structure to make the right decisions and to communicate those "right" decisions in official information to ensure action.	5. An emergency, by its very nature, is characterized by decentralized and pluralistic decision making, so autonomy of decision making should be valued, rather than the centralization of authority.
6. It creates a closed system intended to overcome the inherent weaknesses of "civil" society to deal with important emergencies.	6. An open system of coordinated effort should be created in which the premium is placed on flexibility and initiative among the various social units at the time of the emergency. The goals should be oriented toward problem solving, rather than avoiding chaos.

(From Dynes, 1993. Used with permission)

do in an actual disaster situation; (5) must be based on valid knowledge, including knowledge of how people typically behave in emergencies, knowledge of the hazard itself, and knowledge concerning the resources needed to respond to the disaster event; (6) should focus on general principles while maintaining flexibility; (7) is partly an educational activity;

(8) must overcome resistance; (9) must be tested; and (10) is distinct from disaster management, in that it is impossible to plan for specific problems that will develop when a disaster actually occurs.

In related work, Quarantelli (1988) discussed other important criteria for disaster planning. First, planning must recognize that disasters are qualitatively different (rather than merely quantitatively different) from smaller events such as accidents or routine emergencies. In contrast with these kinds of events, disasters place community systems under extreme stress; responders face new and different demands; and large numbers of often unfamiliar organizational actors (e.g., federal or central government agencies, outside relief organizations, or emergent groups) are involved. Thus planning for disasters cannot be merely an extension of planning for everyday emergencies.

Second, while disaster agents differ from one another and typically require specialized resources, planning efforts should be generic rather than agent-specific, because the same general tasks must be performed regardless of the type of disaster. No matter what type of disaster occurs, for example, there will also be a need for emergency protection, expedient hazard mitigation, population protection, and incident management (see Table 2.3, adapted from Lindell and Perry, 1992, 1996; and Lindell, 1995).

Third, planning is most effective when it is integrated rather than fragmented. Rather than planning independently of one another, the organizations and community sectors responsible for the performance of disaster-related tasks (e.g., medical-care organizations, law-enforcement agencies, fire agencies, local emergency management agencies, and lifeline organizations) should emphasize community preparedness efforts. This principle applies not only to the development of formal disaster plans, but also to disaster exercises, training activities, and other aspects of preparedness.

Disaster researchers have long argued that emergency preparedness is distinct from the development of a written disaster plan (Dynes, Quarantelli, and Kreps, 1981; Lindell and Perry, 1980). It is possible to lack a formal disaster plan and yet be prepared for a disaster because all responding personnel have the knowledge, skills, and equipment for responding to the demands of an incident. Such a situation is most likely in cases involving frequently-encountered minor incidents, such as localized floods and small earthquakes and windstorms, and in slow-onset emergencies. Conversely, it is possible to have a written plan yet be unprepared for emergencies because those who are assigned roles by

TABLE 2.3 Emergency Management Tasks and Activities

EMERGENCY ASSESSMENT
- Threat detection and emergency classification
- Hazard and environmental monitoring
- Population monitoring and assessment
- Damage assessment

EXPEDIENT HAZARD MITIGATION
- Hazard source control
- Impact mitigation

PROTECTIVE RESPONSE
- Protective action selection and population warning
- Protective action implementation
  - Evacuation transportation support
  - Evacuation traffic management
  - Sheltering (in-place protection)
- Impact zone access control and security
- Reception and care of victims
- Search and rescue
- Emergency medical care and morgues
- Hazard exposure control

INCIDENT MANAGEMENT
- Agency notification and mobilization
- Mobilization of emergency facilities and equipment
- Internal direction and control
- External coordination
- Public information
- Administrative and logistic support
- Documentation

---

the emergency operations plan are unaware of them, are insufficiently trained, or lack the resources to perform those roles.

However, although plans are only one element in overall preparedness they do constitute a very important element. First responders, emergency planners, and disaster researchers all contend that Emergency Operations Plans (EOPs) should be derived from a careful analysis of the types of hazards to which a community is vulnerable and an assessment of the community's resources for responding to those hazards. These resources include trained personnel, specialized equipment, support facilities, and financial resources. The purpose of the EOP is to define emergency response functions and allocate responsibility for performing each of them to different community organizations. Although a written plan is an important component of a community's emergency preparedness, a plan is not a step-by-step guide for disaster responders. Step-by-step procedures are important job aids for performing tasks that are infrequent, cognitively complex, and critical to safety, but such procedures should not be confused with emergency plans nor should they be included in those plans.

Instead, a plan should be thought of as having at least two main purposes. First, it provides internal documentation (i.e., within the community) or a "written contract" that reflects all responding organizations' agreements regarding the allocation of emergency response functions, the activation of the emergency response organization, and the direction and control of the response. The second purpose is to serve as a training document. That is, rather than merely being developed and filed, plans should serve as the basis for drills and exercises in which the organizations involved in the planning process are required to carry out their assigned roles in simulated emergencies that resemble those they are likely to encounter. A related training function involves its use as a basis for hazard awareness and education programs for the general public, who need to know about the hazards to which they are vulnerable, what they can expect community organizations to do to protect them when a disaster strikes, and what they must do to protect themselves.

Focusing on preparedness for chemical emergencies, the studies conducted by Lindell and his colleagues on LEPCs (Lindell and Whitney, 1995; Lindell et al., 1996a, 1996b; Whitney and Lindell, 2000) indicate that there are some structures and strategies that are likely to significantly improve the success of at least this type of preparedness network, regardless of context, and equally importantly without significant expense. This finding is consistent with previous studies showing that external constraints can be circumvented to some extent by a superior planning process (Kartez and Lindell, 1987, 1990).

In particular, these researchers found that LEPCs become more effective when they invite representation from agencies and organizations that possess varied knowledge, skills, and interests. Technical materials provided through "vertical diffusion" by federal agencies (e.g., DOT, EPA, and FEMA) also have a positive impact on LEPC effectiveness. Moreover, lateral diffusion of emergency preparedness practices and resources from private industry and neighboring jurisdictions can provide vicarious experience with disaster demands by demonstrating the effectiveness of specific innovations, including plans, procedures, and equipment (Kartez and Lindell, 1987). Other relevant factors affecting LEPC effectiveness include the designation of specialized subcommittees and the use of planning approaches that elicit significant member inputs (e.g., number of members, length and frequency of meetings, high levels of effort and attendance, and low levels of turnover).

The LEPC research also points to the importance of team climate, defined as members' interpretations of features, events, and processes that take place in their work environment. Important dimensions of team climate and individual members' jobs include role stress (role ambiguity, conflict, and overload), intrinsic or extrinsic rewards for emergency planning activities (job challenge and task significance), and characteristics of LEPC leadership (leader goal emphasis and leader support) and of the workgroup itself (workgroup cooperation and team pride). Organizational climate presumably affects LEPC effectiveness because it influences the degree to which members' motivation is aroused, maintained, and directed toward group goals (Lindell and Whitney, 1995). Team climate also is important because it is related to job satisfaction and organizational commitment which, in turn, are related to member participation (effort, attendance, and intentions to remain with the LEPC).

The need for further examination of individual members' perspectives was confirmed by Whitney and Lindell (2000), who discovered that members' organizational commitment was significantly influenced by effective LEPC leadership (the ability to structure team tasks, to communicate clearly, and to show consideration for team members) and by members' perceptions of their own competence. Other factors affecting commitment included identification with LEPC goals (perceived hazard vulnerability and perceived effectiveness of emergency planning) and perceived opportunity for reward (public recognition and personal skill development). In turn, LEPC members' organizational commitment was correlated with their attachment behaviors (attendance, effort, and continued membership in the organization).

Other Questions for Future Research

Although a great deal of progress has been made over the past 25 years in understanding emergency preparedness and its determinants, much remains to be done. Research is needed to more accurately characterize structure of local emergency preparedness networks. What organizations are involved in local disaster planning around the country and how are they linked? What factors are associated with network integration and coordination? Do resources such as funding, personnel, and overall levels of community affluence make for better preparedness, or are other factors more important? To what extent have local emergency managers become professionalized and to what degree are they aware of the state of the art in disaster planning?

Research can also help to identify and evaluate ways of increasing community support for emergency preparedness. Additionally, there is also a need to evaluate alternative methods of interorganizational coordination. Training and job performance aids for emergency responders need to be developed and evaluated. Even critical tasks that are infrequently performed and physically or cognitively complex are quite susceptible to skill decay in the absence of practice. Moreover, the inherently nonroutine character of disasters means that the conditions under which response tasks need to be performed cannot be predicted precisely and or practiced repetitively. Thus, more needs to be known about how to devise training methods that maximize the generalizability of task performance across conditions and are resistant to decay, yet that minimize the time and expense required during initial and refresher training sessions. While Ford and Schmidt (2000) have made a valuable beginning by identifying training problems and solutions that are unique to emergency response training, more work in this area is needed.

A similar need exists for research on the processes by which research findings are disseminated to practitioners. As noted above, many emergency planners and responders believe in disaster myths and engage in ineffective and problematic planning processes despite the fact that many of these problems have been known to disaster researchers and federal agency personnel for at least the past 25 years. Studies of the dissemination process should examine the factors that affect the diffusion of disaster planning innovations—factors that could include local emergency planners' professional training and their status in the community, as well as their educational levels, access to resources, and integration into intra- and extra-community professional networks.

Although the disaster literature is replete with recommendations on how best to undertake disaster planning and how to develop emergency operations plans, these recommendations have largely not been followed up by systematic research. For example, we are unsure about the extent to which emergency planners and disaster managers either know or agree with these principles. Additionally, we have very little information on the extent to which recommendations on how best to prepare for disasters have actually informed local planning efforts.

Most importantly, as has been noted, we have yet to determine in any systematic fashion whether planning efforts based on "correct" emergency management assumptions actually have the intended effect when a disaster occurs. Nor is there much evidence showing linkages between planning assumptions and actual organizational performance in disaster situations. In a study of the community response following the 1980 Mt. St. Helens volcanic eruption, for example, Kartez (1984) showed that local governments did employ strategies that were consistent with research recommendations. However, their use of adaptive measures was the result of improvisation during the disaster, not prior planning. Generally speaking, although evidence has been gleaned from case studies and research on small samples, nearly three decades after many widely-accepted ideas on preparedness planning were first introduced, we have yet to develop solid empirical evidence on the extent to which they actually improve the ability to respond or the factors that account for successful implementation.

Finally, as the focus moves from households through organizations, communities, states, and nations, progressively less research exists. The amount we know is, in other words, inversely related to the level of analysis studied. This suggests a need for more preparedness research that begins at the "top"—with cross-national and national-level research. Obviously it is extremely important to understand which households prepare for disasters and why. But those questions are equally relevant for organizations, communities, and higher governmental levels.

CHAPTER THREE

Moving into Action: Individual and Group Behavior in Disasters

ISASTER RESPONSE ACTIVITIES consist of actions taken at the time a disaster strikes that are intended to reduce threats to life safety, to care for victims, and to contain secondary hazards and community losses. These actions may be initiated before disaster impact if there is adequate forewarning but usually can take place only after impact in the case of agents such as earthquakes, which occur without warning. As outlined in Table 2.3 (see previous chapter), response measures include *population protection* activities, such as warning, evacuation, search and rescue, and the provision of emergency shelter and emergency medical care. They also include *expedient hazard mitigation* actions, such as installing temporary hurricane shutters, sandbagging flooded rivers, and controlling the secondary impacts that result from disasters, such as earthquake-induced fires.

As Dynes, Quarantelli, and Kreps (1981) long ago observed, emergency response activities must address not only the "agent-generated" demands such as those described in the previous paragraph, but also "response-generated" demands—that is, they must take steps to ensure effective management of the disaster event. These response-activities can be further categorized into tasks centering on *emergency assessment* and those concerned with *incident management*. Emergency assessment tasks include ongoing hazard monitoring and the assessment of both physical damage and impacts on at-risk populations. Incident management encompasses activities associated with the notification and mobilization of responding organizations, intra- and interorganizational coordination, and intergovernmental relations during the emergency period.

The emergency response phase, which was the original focus of disaster studies when the field began to develop, has also been the most-studied phase of the disaster cycle. Significantly more is known about response, for example, than about mitigation or recovery. This is partly because disaster research continues to be driven to a large degree by the study of specific disaster events.

Many of the comments that were made about research on emergency preparedness in the previous chapter also apply to research on emergency response. Conceptual frameworks, research designs, and the variables included in analyses vary widely across studies, making generalization difficult. Some topics, such as household response to disaster warnings and population protection actions generally, have been studied quite extensively, while other equally significant ones, such as post-disaster sheltering, have received little emphasis. And studies that focus on the more micro-social units of analysis like the household are much more common than studies at the macro-social level.

In this chapter, we will first discuss research findings on household responses in disasters, and then examine the response activities of other groups, such as emergent search and rescue groups and disaster volunteers. We will review what research has shown on such topics as how people respond to warnings, including how evacuation decisions are made; patterns of emergency shelter and short-term housing; public involvement in search and rescue; and volunteer activities during the emergency response period. As in Chapter Two, this chapter also highlights important questions the field has yet to adequately address.

Even in a lengthy volume such as this, summarizing and evaluating all the work that has been done on emergency response over the last 25 years is impossible. Instead, the sections below contain general overviews of the research that has been conducted and short descriptions of selected research projects. Studies were chosen for discussion because of their importance and because they illustrate the variation that exists in approaches to studying disaster response. The literature on response contains a large number of small, specialized, and single-case descriptive studies. Rather than attempting to recapitulate all that work, our review emphasizes large-scale studies, projects that consider multiple cases, and exemplary and ground-breaking research.

HOUSEHOLD EMERGENCY RESPONSE

Warning Response Research

Most of the research on household emergency response has focused on warning receipt and protective response activities. More specifically, this research has examined such issues as the sources and channels from which people receive warnings, the credibility of those sources, when people receive warnings, and the degree to which they pass on warnings and seek further information from friends, relatives, neighbors, and coworkers.

For many years, research on protective response behaviors mainly focused on compliance with authorities' evacuation recommendations. Much of the research conducted on this topic prior to the mid-1980s has been reviewed previously by Vogt and Sorensen (1987) and Sorensen, Vogt, and Mileti (1987). A diverse collection of articles on U.S. evacuation research appeared in a 1991 special issue of the *International Journal of Mass Emergencies and Disasters* (Volume 9, No. 2), and another overview was provided by Sorensen (1993). However, it should be kept in mind that not all protective action involves evacuation and relocation elsewhere. In some cases, sheltering-in-place and "vertical evacuation" (discussed below) can be equally effective alternatives.

Research on household emergency response has identified a series of social, social-psychological, and cognitive processes that shape the actions of threatened populations. Social-structural factors such as degree of community integration help to determine who receives a warning and when they receive it, as well as the channels through which the warning is received. Social-psychological processes affect how those who receive the warning assess both the source and the warning message. Cognitive processes influence how this information is handled as people try to reach decisions about how to respond to the threat. As will be described in more detail below, past research has yielded a number of findings that have been replicated by different researchers using different methods (e.g., case studies versus surveys) across a variety of different types of hazard agents (e.g., hurricanes, tornadoes, floods, and various technological hazards).

The research on household emergency response has a number of limitations. Perhaps the most significant drawback is that studies have typically been conducted on events within a single jurisdiction, thus confounding characteristics of the disaster event with characteristics of the community. Those attempting to review research findings across studies thus must attempt to separate out the effects of variations between hazard agents (e.g., floods versus hurricanes), between different incidents involving a particular hazard agent (e.g., among all hurricanes), and across different communities and regions.

As is the case with research on household emergency preparedness, studies of household emergency response have been marked by differences in researchers' theoretical perspectives, as well as in the operationalizations of the variables on which they do agree. For example, hazard experience and perceived risk—to name only two variables that are considered important for predicting how households respond in emergencies—have been measured in almost as many different ways as there are researchers. This is a particularly severe problem because of the literature's large number of "one-shot" investigations that have been conducted on specific disasters using idiosyncratic approaches that all too often have not taken previous research into account. This problem is often compounded by incomplete reporting of results, especially the failure to report intercorrelations among all variables investigated in different studies, which makes it impossible for later investigators to examine the validity of hypotheses different from those tested by the original investigators. Finally, though the discussion below indicates that many variables have been repeatedly found to have statistically significant effects, little attention has been given to the overall variance in behavior that is accounted for by the predictor variables or to the strength of predictions involving a given variable over different studies.

Although rarely acknowledged explicitly, most warning response studies employ some variant of the Source-Channel-Message-Receiver-Effect-Feedback communication model (Lasswell, 1948). Specifically, information about an actual or potential disaster can come from physical cues or from social sources such as authorities, news media, and informal groups. The information can be transmitted face-to-face, or through different technological channels (print or electronic) to different demographic segments of the community, producing a range of psychological and behavioral effects. The research assumes that the effects on the recipient take place in a sequence of stages, including exposure to the information, attention to it, comprehension of its meaning, and acceptance of its accuracy and relevance for the receiver. This information process-ing yields two important types of psychological effects—cognitive reac-tions such as perceptions of threat and of alternative protective actions, and affective responses such as fear. In turn, these psychological effects lead to behavioral consequences, which can range from the continuation of normal activities to undertaking personal- and property-protection measures. The loop is then closed when recipients obtain feedback, ei-ther by seeking additional information or by observing the effects of their actions.

With respect to passage through the stages of this process, investiga-tors consistently have found that recipients initially disbelieve warnings, which instigates attempts to confirm the threat from other sources. Warn-ing disbelief sometimes has been confused with the psychodynamic term "denial," but the latter term generally is not appropriate because it refers to a refusal to acknowledge an unambiguous and immediate threat. In contrast, in many cases disbelief is an entirely logical reaction, in that warnings generally involve improbable events and are issued in unusual, confusing situations. A warning message is most likely to motivate timely and effective action if it creates a perception of the threat as being certain to occur and as having severe and immediate consequences for recipi-ents. Moreover, protective action is more likely to be undertaken if the warning describes (or leads recipients to recall) a protective action that is effective, but at the same time doesn't involve large monetary costs, time and effort requirements, or other barriers, such as the need for special-ized knowledge or cooperation with others (Dynes and Quarantelli, 1976; Carter et al., 1977; Perry et al., 1981; Houts et al., 1984; Lindell and Perry, 1992).

Warning Dissemination

As Lindell and Perry (1992) note, warning systems seldom operate smoothly, for various reasons. Sources, channels, and messages differ in their effects on recipients, and recipients differ from one another in terms of the sources, channels, and messages to which they will be most recep-tive. To further complicate matters, the mechanisms through which warn-ings are issued differ, and there appears to be no universally preferable strategy for conveying warnings. For example, in the U.S., a face-to-face warning by a uniformed officer is probably the most credible warning mechanism for the majority of the population, but that method has the disadvantage of being very slow and labor intensive. Sirens achieve rapid dissemination, but they only convey the general idea that something dangerous is taking place, as opposed to a specific warning about what is happening, where it is happening, and what actions people should take to protect themselves.

Other warning mechanisms—such as route alert loudspeakers, tone alert radios, and commercial telephone, radio, and television—vary along a number of dimensions, including precision of dissemination, their ability to get people's attention as they go about their normal activities, the specificity of the message that can be conveyed, susceptibility of the message to distortion, the rate of dissemination over time, receiver and sender requirements, the ability to verify warning messages upon receipt, and initial and ongoing operating costs (Lindell and Perry, 1987; Sorensen and Mileti, 1987). From a research standpoint, these variations make it quite difficult to generalize across studies involving warning response. From a practical standpoint, they make the design of warning systems and the issuing of warnings very challenging.

Another significant problem with the dissemination of information to the public in emergencies involves the conflicts that can develop among information sources in their assessments of the threat or their recommendations for protective action. There are a number of different ways in which conflict can arise among warning messages. The first such conflict can be the result of overlapping broadcast areas, which causes information that is accurate for one geographic area to be received in another area where it is inaccurate. Another basis of conflict stems from differences in the timeliness with which sources update changing information about the situation. That is, information that was correct at one time may be received by some people at a later time when it is no longer accurate. Conflict can also arise from differences among sources in terms of their assessments of the situation. Specifically, one source may have more accurate information or more expertise for processing that information than another, but these differences may not be recognized by those to whom the warnings are disseminated. Further complicating the matter, the social cues people obtain through observing the actions of others who are responding in an emergency can either enhance or undercut the recommendations made by authorities. Research has also repeatedly confirmed the presence of other response conflicts, such as the need to ensure the safety of other family members and to evacuate as a household unit, even if that means evacuation is delayed (Drabek and Boggs, 1968; Drabek, 1983a).

Previous research also has identified a number of myths that exist about disaster behavior that influence the responses of both emergency managers and the general public. These erroneous expectations, often mutually contradictory, include the assumption that individuals will respond with docile obedience to authority, that they will be immobilized due to emotional shock and unable to respond, and that they will engage in panic flight when warned of an impending disaster. Other myths include fears about a lack of local resources for response, low community morale, role abandonment by emergency personnel, and looting and social chaos (Quarantelli and Dynes, 1972; Wenger et al., 1975; Kreps, 1991; Lindell and Perry, 1992; Fischer, 1998). Erroneous beliefs about how people behave in disaster situations appear to be quite widespread among both the public and disaster management officials. They also tend to persist even in the face of personal experiences in actual emergencies that contradict disaster myths (Fischer, 1998). These kinds of beliefs can undermine the warning process in various ways. For example, officials may delay issuing warning messages because of a concern that doing so might create panic, or residents may refuse to leave when a warning is issued or return before it is safe to do so out of fear that their homes will be looted.

Warning Responses Other Than Evacuation

For many years, research equated protective response with evacuation and relocation, but over the past two decades researchers have begun to examine other types of protective actions that can be undertaken by threatened populations. Sheltering in place, as opposed to evacuating, has long been recognized as the appropriate response to tornadoes, but this form of self-protection has only received consideration in the case of other hazards since it began to be advocated as an appropriate response in emergencies involving nuclear and chemical hazards. Similarly, problems in evacuating densely-populated coastlines on the Atlantic Ocean and the Gulf of Mexico have stimulated consideration of sheltering in place as a protective response for hurricanes.

Berke (in Ruch et al., 1991) addressed the potential usefulness of sheltering in place—for example, relocating to upper floors in multistory buildings—as a strategy for addressing the problem of providing emergency shelter. There are two possible uses of sheltering in place: as a refuge of last resort, and as a "planned supplement to horizontal evacuation" (Ruch et al., 1991: 2). Evacuating "internally" or "vertically" may reduce traffic volume in affected areas and in some cases avoid or drastically reduce the need for evacuation.

Differential Responses to Protective Action Recommendations

Recent literature raises some important questions about the extent to which community residents comply with warning messages involving different types of disaster agents and whether the public responds in atypical ways when particular types of disaster agents, such as nuclear hazards, are involved. On the one hand, research has consistently found a pattern of under-response to threat, characterized by warning responses that are too slow and incomplete in the risk area. As noted above, disbelief is a common initial response to warning messages, and many people appear determined to remain in harm's way when disaster strikes despite clear and specific warnings. On the other hand, research on evacuation during the Three Mile Island nuclear plant emergency found evidence of an "evacuation shadow." Specifically, Zeigler, Brunn, and Johnson (1981) found that there was movement out of the area by people living *outside* the risk zone designated by the governor's protective action order (Zeigler, Brunn, and Johnson, 1981). Other studies of Three Mile Island and of radiation hazards more generally concluded that an evacuation shadow probably arose because people judged by authorities not to be at risk nevertheless came to define themselves as in danger because of confusing and conflicting information from authorities, geographic proximity to the plant, and similarity to demographic groups targeted in the warning messages (Houts et al., 1984; Lindell and Perry, 1983; Lindell and Barnes, 1986). In other words, while in some crisis situations people within identified hazardous areas show a marked tendency not to move when they're told, in other kinds of emergencies people who are outside designated risk areas move even though they are not told to. The latter situation can add to traffic congestion and cause confusion about which areas are safe and which are not.

Some researchers (e.g., Zeigler and Johnson, 1984) have concluded that the evacuation shadow phenomenon is specific to radiological hazards. However, other examples of this phenomenon have been documented in connection with a chlorine tank car derailment at Mississauga, Ontario, and the eruption of the Mt. St. Helens volcano (Lindell and Perry, 1992). Moreover, Gladwin and Peacock (1997) also found evidence of a significant shadow effect in their study on Hurricane Andrew; about one-fifth of the households in coastal areas that were not under an evacuation warning left anyway. Given these kinds of findings, one might wonder why the evacuation shadow phenomenon took so long to discover. The answer appears to be that early studies on natural hazards failed to find evidence of an evacuation shadow because of the methodology they used, which generally only sampled respondents from within the areas covered by warnings.

Other Research Issues

The concept of risk perception has long played a central role in explaining why people respond the way they do to disaster warnings, just as it helps explain preparedness and other self-protective actions taken before disasters strike. Unfortunately, there have been few if any advances in our understanding of the mechanisms of risk perception in the disaster context. Operationalizations of the concept tend to be idiosyncratic. Risk perception is measured in a variety of ways: as the perceived likelihood of a particular type of event, such as an earthquake; as the perceived magnitude of an event; as expectations about the severity of its impacts on the community; and as expectations about the personal threat posed by the hazard. Some recent studies have adopted a multiple operationalization strategy that asks respondents a variety of questions about the perceived risk (e.g., Mileti and Fitzpatrick, 1993), while others have examined respondents' perceptions of multiple attributes of the risks of different hazard agents (Lindell, 1994b). Further studies of these types are needed to identify those conceptualizations of risk that are most defensible theoretically and most strongly correlated with behavior.

Additionally, relatively little attention has been given to studying the effects of personality characteristics on emergency response, and that research has been limited to generalized expectancies for internal versus external fate control. The results of the most widely known locus-of-control study (Sims and Bauman, 1972) proved controversial, and its conclusions could not be replicated in other investigations. Other research (Wood and Bandura, 1989) suggests that task-specific self-efficacy is more relevant to performance of a specific action than are generalized expectancies.

One important research need is for more accurate modeling of protective action decisions such as evacuation and sheltering in place. Various empirical and conceptual models of evacuation behavior have been proposed. Sorensen and Richardson (1984) developed a causal model that attempted to explain evacuation decisions made following the 1979 nuclear power plant accident at Three Mile Island (see Figure 3.1). In this model, which subsequently was supported by studies of other emergencies, the decision to evacuate is characterized as the result of the direct and indirect influence of ten different factors: hazard characteristics, situational constraints, perceived threat, the information provided, concern over risk, coping ability, attitudes towards risk managers, demographic characteristics, risk sensitivity, and social ties.

An evacuation model formulated by Quarantelli (1984), which is more of a conceptual or analytic model, suggests that five sets of factors are important for understanding evacuation behavior: the community context, which includes available resources and existing preparedness planning efforts; threat conditions (e.g., characteristics of the disaster agent); how the threat is defined by residents; resultant response-related social processes at the community and organizational levels, such as efforts at communication and task allocation; patterns of behavior, such as the issuance of warnings, the evacuation activities themselves, and sheltering behavior; and the impacts or consequences those actions have for future preparedness and response activities.

Perry, Lindell, and their colleagues (Perry, Lindell, and Greene, 1981; Lindell and Barnes, 1986; Lindell and Perry, 1992) have developed and tested a model of protective response that is based on behavioral decision

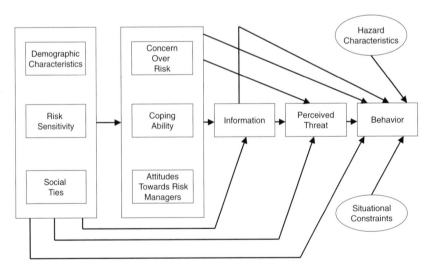

Figure 3.1 A General Model of Evacuation Behavior (From Sorensen, Vogt, and Mileti, 1987)

theory. The "protective action decision model," which is shown in Figure 3.2, is based upon a conceptual integration of emergent norm theory (Turner and Killian, 1987) and general systems theory. This model can be represented by a decision tree consisting of four questions that people must address when deciding whether or not to comply with a warning message. The primary decision nodes represent determinations of the believability of the threat, whether it is even technically possible to be protected from the threat, whether prevailing conditions allow one to pursue protective options, and whether undertaking protection significantly reduces negative outcomes. On the right side of the flow chart, characteristics of the decision-maker, situational (environmental) factors, and social factors are identified, and the points at which they impinge on the different decision points in the model are indicated. The expanded model has been tested and found to explain statistically significant amounts of variance in protective response to floods, volcanic eruptions, and hazardous materials accidents. The model also was modified and successfully used to predict the adoption of longer-term mitigation measures by citizens facing volcano and earthquake threats (Perry and Lindell, 1990a, 1990c; Lindell and Whitney, 2000).

The studies discussed above suggest similar frameworks for conceptualizing the evacuation process. An evacuation order, no matter how clear, scientifically based, specific, urgent, and authoritative, neverthe-

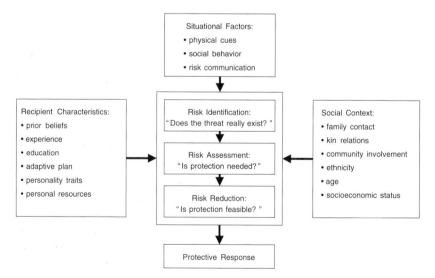

Figure 3.2 The Protective Action Decision Model (Based on Lindell and Perry, 1992)

less is embedded in a particular social context and influenced by social-structural factors and ongoing social routines. Evacuation decisions are affected by observable cues in the environment, such as wind and rain, as well as by message and warning system characteristics. Other influences include the psychological, sociodemographic, and sociocultural characteristics and past experiences of the individuals and groups that receive disaster warnings. These factors are shaped in turn by the broader community context, as well as by the immediate context in which the message is received and evaluated—for example, whether household members are able to account for one another and evacuate together, whether it is possible to confirm the evacuation warning through personal contacts, and whether evacuation is possible from a practical standpoint.

It is important to note that although much is known about factors affecting evacuation decisions at the household level, current explanatory models only account for 50 percent of variance in warning response at best (e.g., Perry, Lindell, and Greene, 1981). Similar uncertainties can be found in estimates of the time component of protective actions. Although warning times are determined by the specific warning mechanisms existing in a given community, and evacuation travel times are significantly influenced by the characteristics of local road networks (see Lindell and Perry, 1992), the time households need in order to prepare to leave are substantially determined by social and psychological processes that have not been well defined. In particular, there is a need for further clarification of vulnerable populations' perception of threat and protective actions, as well as factors affecting their perceptions of information sources. Evidence from a variety of investigations indicates that information sources are perceived in terms of their expertise, trustworthiness, and attractiveness, and that these characteristics are affected by a source's credentials, acceptance by other sources of known credibility, and past interactions with the receiver. The degree of hazard knowledge attributed to different sources also varies from one disaster agent to another (Lindell and Perry, 1992). Given the importance of source credibility in determining warning response, further investigation of this area is warranted.

As noted above, sheltering in place is increasingly being seen as a protective action that can be used in response to various hazards, including tornadoes, hurricanes, and hazardous chemical releases. However, recommending that people shelter in place will not work unless people are convinced that doing so provides adequate protection. Accordingly, Ruch (1991) conducted two studies on willingness to shelter in place during a hurricane. His data led him to conclude that "more people perceive that vertical shelter is safer than traditional shelter," but that at the same time "people have a strong aversion to vertical shelters" (1991: 7). Moreover, the idea of sheltering in place may be more appealing to urban populations, or to people who live in particular regions of the country. Berke (1991) for example has noted that Texans' strong aversion to any regulation that seems to jeopardize private property rights generates resistance against such measures, while Florida residents seem more receptive to sheltering in place. Studies like these are a promising start, but since they were conducted there appear to have been no other studies on sheltering in place as an option during hurricanes.

The research on sheltering in place as a protective action in hazardous materials emergencies is similarly limited. Lindell and Perry (1992) reported data on the perceived efficacy of sheltering in place and on intentions to evacuate rather than choosing that option for three hypothetical hazards: a volcanic eruption of Mt. St. Helens, a chlorine release from a railroad tank car, and an accident at a nuclear power plant. Survey respondents thought that sheltering in place would be significantly more effective for a volcanic eruption than for the chlorine release or the nuclear accident. Accordingly, their estimates of the likelihood of evacuating rather than sheltering in place if the latter were recommended were significantly lower for the volcanic eruption than for the other two threats.

Other data indicate that people see evacuation as providing more protection than sheltering in place, but that the latter is seen as posing fewer barriers in terms of the time, effort, and financial investment involved (Lindell and Perry, 1992). Clearly, more research on warning recipients' behavior in actual emergency situations—as opposed to the hypothetical ones just discussed—is needed before conclusions can be drawn about sheltering in place as a population protection option.

Gaps in the Literature: Protective Responses by Travelers and Transient Populations

Virtually all research on warning response and evacuation has focused on households at their place of residence—or at least in the local communities where they live—and the ways in which they receive and act on disaster warnings. But how do people react when they are not at home, or even in their home communities, when disaster strikes? Do families on vacation in Hawaii, for example, behave differently from the way they would at home when a hurricane warning is broadcast? Are they more or less likely to receive, believe, and act on warning messages? What about people who don't live in a settled location to begin with? Are their responses similar to or different from those of more established community residents? These are the kinds of questions Drabek addressed in his work on evacuation behavior among tourists and other transient populations (1996). For that study, Drabek interviewed over 500 tourists, 83 business travelers, and dozens of migrant workers and homeless people, as well as lodging industry representatives and local public officials.

Drabek found that the warning responses and evacuation behaviors of these groups differ in several important ways from those of people who receive disaster warning messages in their homes or home communities. For example, tourists and transients are less likely to receive warning messages from the mass media. Instead, their warning information is more likely to come from other sources, such as hotel employees. As a consequence, these groups are likely to receive warnings later than residential populations and to have less time to act on them. It also appears that tourists and transients react to warnings in more extreme ways than community residents, in that they tend to either discount warnings completely or evacuate immediately. Moreover, more settled populations typically turn first to relatives and friends for shelter when they are forced to evacuate. The groups studied by Drabek, which typically lack that option, were more likely to rely on public shelters; some sought out other lodging establishments or simply left the endangered area and went back home. Despite these differences, it appears that tourists and transients respond to warning messages in ways that resemble those of residential populations; for example, they seek to confirm warning messages with others and talk with others about what they should do next.

The Drabek study, which to our knowledge is unique in the literature, calls attention to the fact that at any given time a risk area is likely to contain substantial numbers of people who are actually from somewhere else, as well as people who have no permanent residence at all. Such transient populations will tend to lack knowledge of the risk area and of suitable evacuation routes and destinations, and they will have less access to warning information. Unlike community residents who resist evacuating because of their desire to protect their property either from disaster damage or from the imagined threat posed by looters, travelers have little motivation to remain behind when told to leave. At the same time, they will not necessarily be reached or influenced by warning messages targeting the general public. In light of these factors, transients' responses to protective action recommendations are likely to differ significantly from those of local residents.

Practical Applications of Household Protective Response Research: Incentives for Warning Compliance

In addition to the substantial amount of research on household warning response, a number of studies have obtained feedback from citizens on actions emergency managers might take in order to produce higher levels of voluntary warning compliance. Findings from such studies can give emergency managers research-based guidance on how to create incentives for warning compliance—i.e., procedures or provisions that will encourage threatened community residents to follow warning instructions (Perry, 1979b; Kartez, 1984). Based on empirical studies, incentives for warning compliance should address five areas of concern: the need for warning confirmation; transportation support; congregate care; family communications; and property protection. In the following section, we discuss how incentives can be developed to address needs in each of these areas.

With respect to warning confirmation, as noted above, research shows that the receipt of a warning initiates a process of information-seeking that aims at confirming warning accuracy and obtaining guidance about what to do next. To address this need, telephone warning hotlines can be established and information about the hotlines can be given in the warning message, which ideally repeats information that has already been disseminated to the public during hazard awareness programs conducted prior to disaster impact. Hotline systems can also perform a rumor-control function during disasters. Further, since confirming information can be standardized through the use of recorded messages, hotlines can minimize problems that routinely arise when residents receive contradictory or conflicting information. The meteoric rise of the Internet as a means of communication also raises the potential for web-based hotlines and information dissemination. In many communities, the Internet is currently being used in the same manner as a telephone hotline, to provide information of all kinds, including warning information, prior to and after disaster impact.

There are, of course, issues that need to be considered in connection with telephone and Internet-based information hotlines. Foremost among them is the potential for system overload. Research dating back as far as

the 1950s (Fritz and Mathewson, 1957) documents telephone convergence as a major problem in disasters, and for decades disaster planning
handbooks have recommended telling people not to use the phone in
disasters (Healy, 1969). Because of tremendous proliferation of cellular
telephones and the growth in the number of households that are connected to the Internet, more people than ever are in a position to access
warning and other emergency information almost instantaneously. At
the same time, this raises the possibility that the demand for information
in a disaster situation will exceed any system's capacity to provide it.
Indeed, the experience in recent California earthquakes indicates that
cells and cellular phones become overloaded almost as readily as the
older switching and circuits (Lindell and Perry, 1996). Clearly emergency
managers need to develop a variety of communications links using such
diverse media as television, radio, conventional telephone lines, cell
phones, faxes, and e-mail and Internet communications to address the
public's need for information. The shift from the Emergency Broadcast System to the digitally-based Emergency Alert System opens up
additional possibilities for the dissemination of emergency-relevant
information.

Research has also identified a number of transportation issues that
influence protective response, particularly when evacuations are involved
(Urbanik et al., 1980; Sorensen and Rogers, 1988; Lindell and Perry,
1992). The first of these centers on the provision of evacuation transportation assistance while the second involves the development and dissemination of traffic management plans. With regard to evacuation transportation assistance, the proportion of U.S. residents without access to cars
is relatively small but varies by region, with household vehicle ownership
being lower in eastern urban areas than in rural western regions. Un-
surprisingly, lack of access to a vehicle (especially a reliable one) can also
be a serious impediment to evacuation compliance for members of lower-
income groups. In many cases, those who do not own vehicles will obtain evacuation transportation assistance from the same people who routinely help them on a daily basis. However, emergency managers should
still be aware of which groups are most likely to have problems with
emergency transportation, and they must not automatically assume that
informal assistance will be forthcoming. Systematizing and publicizing
the availability of evacuation transportation support and providing information on staging areas and pickup routes would likely enhance
evacuation compliance, as would publicly-supplied high-occupancy ve-

hicles, which would have the benefit of reducing traffic congestion during evacuations.

With respect to traffic management plans, it is important to recognize that households lacking evacuation plans are less likely to evacuate and slower to act when they do evacuate, and that they also have a tendency to evacuate to even more dangerous locations rather than safer ones (Lachman, Tatsuoka, and Bonk, 1961; Simpson, 1980). And for the majority of evacuating households in the U.S. who will use their own vehicles, emergency managers must anticipate traffic management problems such as overcrowded evacuation routes, accidents, and vehicle breakdowns. It is possible to use computer-based planning models to evaluate evacuation plans for areas with large populations and complex road networks (Urbanik, 1994). However, simplified analyses using manual calculations can also produce satisfactory evaluations of evacuation route restrictions in smaller areas (Lindell, 1995). Once emergency managers have developed and assessed their evacuation management plans, they should disseminate important traffic management information such as risk area maps and evacuation route information to community residents as part of their overall community planning efforts. This kind of information, which already has been prepared in connection with emergency plans for nuclear power plants (Lindell and Perry, 1996), can also be used for other types of hazardous facilities, as well as for natural hazards such as hurricanes and riverine floods. Residents of risk areas can be encouraged to make contact with friends or relatives in safe areas and to arrange in advance to stay with them if evacuation is required. Such arrangements would permit evacuees to depart as soon as a warning is issued.

The provision of public congregate care facilities can also be used as an incentive for evacuation. Research indicates that evacuees tend to underutilize public congregate care facilities (Moore et al., 1963; Perry, Lindell, and Greene, 1981), which are commonly viewed as accommodations of last resort, and that such facilities rarely attract more than one-fourth of those who evacuate. In most cases, congregate care utilization is likely to be in the range of 5–15 percent (Lindell et al., 1985), but that proportion can vary as a function of population, event, and community characteristics. Specifically, evacuees are more likely to rely on congregate care facilities if they are less integrated into the community (reducing the number of friends and relatives with whom to stay); have lower incomes (reducing their ability to afford hotel or motel accommodations); and rely on public transportation (reducing their ability to reach alternative sources of shelter). Drabek's (1986) review indicates that higher levels of public congregate care use are associated with agent characteristics such as rapid onset, little forewarning, large scope of impact, high levels of destruction, and short duration, and relatedly with agent type, with chemical and nuclear agents stimulating more public shelter use. Situational factors encouraging public shelter use include nighttime evacuations and bad weather, both of which discourage people from traveling longer distances in search of suitable alternative shelter. Community characteristics that are associated with greater congregate shelter use include isolation from other communities and high levels of preparedness. Such facilities are also more likely to be used when an entire community has to be evacuated. Finally, even in disasters in which most evacuees ultimately find shelter with friends and relatives or in hotels, many residents will still depend on public facilities, at least initially.

A related issue concerns the question of how people become aware of different sheltering options. Warning messages are one such source of information. Most evacuees identify an evacuation destination by making the first contact themselves during the warning period. However, in a few cases evacuees reported that they sought out a particular place because they had identified it in their family disaster plans. Additionally, three decades ago Drabek (1969) documented a pattern he termed "evacuation by invitation," in which people in an endangered area are contacted by relatives and friends who invite them to come to their homes. This phenomenon has been noted in subsequent research as well (Perry, Lindell, and Greene, 1981; Perry, 1985).

These findings have several implications for evacuation and shelter planning. First, when given an opportunity to choose, people tend to avoid public congregate care facilities in favor of staying with people they know. Second, although people can find out about congregate care facilities through warning messages, it is better if they are informed about them in advance. The literature also argues strongly for the use of flexible plans for the emergency sheltering of evacuees, such as the use of reception centers. At such facilities, evacuating families can register with authorities, obtain additional information, and then leave their names and information about where they will be staying.

Registration with authorities is especially important because of the strong influence household members exert on one another under disaster threat conditions. Households tend to evacuate as units (Drabek and Boggs, 1968; Drabek and Key, 1984), and the separation of household members often involves anxiety and prompts attempts to reunite, sometimes by returning to previously evacuated areas. It may be the case, however, that uniting households is not necessarily as important as simply being able to provide information on the whereabouts of household members (Hans and Sell, 1974; Haas, Cochrane, and Eddy, 1977). These findings suggest that evacuation compliance can be encouraged through mechanisms such as family message centers where evacuees can obtain information about other household members. If such information centers are included as an element in reception centers and shelters, people may be more willing to heed evacuation warnings.

Since looting is rare in natural disasters, it is unnecessary and perhaps even a poor use of personnel to deploy large numbers of police or troops to guard evacuated areas (Quarantelli and Dynes, 1970a). Symbolic security measures generally are sufficient to protect property in the majority of cases (Dynes, Quarantelli, and Kreps, 1981). Nevertheless, residents' perceptions of security problems do have a significant influence on evacuation behavior, and thus they need to be taken into account. To encourage evacuation compliance, emergency management officials should communicate to the public how they intend to ensure security when an area is evacuated. If officials are seen as having the situation under control, people will be less likely to stay behind to protect their property.

The incentives suggested here are examples of practical measures that can be developed from empirical research on household emergency response. These recommendations follow directly from the basic idea that emergency plans will be more effective if they are based upon knowledge of how people actually behave in emergencies and if they are geared toward overcoming actual and perceived barriers to taking self-protective action (Dynes, Quarantelli, and Kreps, 1981).

Many additional topics warrant further study. For example, while it seems intuitive, it is in fact not clear whether and how prior disaster experience influences household emergency response actions. Evidence has also been inconsistent on the effects of prior personal experience with hazards on risk perception and response. In his summary of hurricane evacuation studies, for example, Baker (1991) concluded that none of the measures of previous hazard experience in the literature were significantly related to hurricane risk perception. The finding that risk perception is unrelated to experience is so completely counter-intuitive (not to mention inconsistent with 100 years of psychological research on human learning) that it suggests a basic defect in our conceptualization of the variables and processes involved. Thus, research is needed on the cognitive mechanisms by which experience with low probability/high consequence events is interpreted, remembered, and retrieved when needed to support later decisions. Such research on natural hazards should be informed by more general theories of social cognition, especially work on judgmental heuristic biases (see for example Tversky and Kahneman, 1973, 1981; Sherman and Corty, 1984; Kunreuther, 1992). As Baker (1991) has noted, there are many different ways of operationalizing both constructs—personal experience and risk perception—and more needs to be known about which operationalizations best predict people's behavior.

Research is also needed on the effects of hazard awareness and education programs on risk perception and protective response. Some initial typologies of these types of programs have been developed (Sorensen and Mileti, 1987), and the psychological mechanisms by which they might operate have been identified (Lindell and Perry, 1992), but empirical evaluations of this work remain to be done. Given the low levels of both household and business preparedness for disasters, there is a significant need for research that can aid in the development of effective public education strategies.

EMERGENCY SHELTER AND HOUSING

Much of what we know about post-disaster sheltering and housing comes from research done in the last 15 years. The process remains significantly understudied, and little research has looked at post-disaster housing patterns across social classes, racial/ethnic groups, and family types. A fuller understanding is also needed of how sheltering and housing are experienced and undertaken at individual, group, organization, and community levels and of the impact organizations and groups from outside stricken communities can have on the process.

The literature makes it clear that post-disaster sheltering and housing encompass both physical and social processes. In the first effort to develop a taxonomy of those processes Quarantelli (1982b, 1982c) conceptualized shelter and housing activities as involving four stages: emergency sheltering, temporary sheltering, temporary housing, and permanent housing. Emergency sheltering, which takes place in the immediate pre- and post-impact periods, is spontaneous and based on expediency. Temporary sheltering, provided formally in the U.S. by organizations such as the Red Cross and informally by friends, relatives, and neighbors, requires the provision of food, sleeping facilities, and other ser-

Rocky Mount, North Carolina — These manufactured homes housed more than 300 families flooded out by Hurricane Floyd.

vices. Although not intended to be anything but brief, temporary shelter may actually last for weeks for disaster victims who lack more satisfactory alternatives. Most emergency preparedness focuses on this stage of the housing process.

Temporary housing involves the reestablishment of household routines in alternative living arrangements, but with the understanding that the household is still waiting for more permanent housing. Very little is known about how households fare in this stage of the rehousing process. However, research does suggest that temporary housing can turn into permanent housing in some circumstances (Bolin, 1994). Permanent housing—the stage for which planning is most lacking—consists of the housing arrangements in which victims ultimately find themselves, whether intended or not. Our current discussion emphasizes emergency and temporary shelter, since those are the forms of sheltering that are most common during impact and the immediate post-disaster response period.

Research suggests that passage through the four sheltering and housing stages is affected by many factors, including existing housing and disaster assistance policies, governmental decisions in specific disaster situations, and cultural practices related to shelter and housing (Bolin, 1982, 1998; Bolton, Liebow, and Olson, 1992; Phillips, 1993; Phillips, Garza, and Neal, 1994; Neal and Phillips, 1995; Phillips, 1998). Just as is the case with respect to other disaster-related behaviors and activities, pre-disaster conditions exert a strong effect on post-disaster housing arrangements. Those conditions include the vulnerability of the housing stock to disaster-related damage, the nature and extent of pre-disaster planning, interorganizational mobilization and communication, pre-impact community conflict, resource and power differentials, and both victim and community socio-demographic characteristics (Quarantelli, 1982b; 1982c; Oliver-Smith, 1990; BAREPP/NCEER, 1992). As noted above in our discussion of evacuation, pre-disaster social ties influence where people go when they are forced to flee their homes in an emergency. Those who have small or weak social networks—for example, few friends or family members on which they can rely—are more likely to use publicly-operated shelters, while those with stronger and more extensive social ties have family members and friends upon whom they can call for help. Pre-existing social inequalities, including differences in income and household resources, ability to own a home, access to insurance, and access to affordable housing also have a significant impact on housing options following disasters.

Studies on the provision of temporary shelter following the 1989 Loma Prieta earthquake illustrate this continuity principle and shed light on the various social-structural factors that influence post-disaster housing patterns. Consistent with what research has found generally, the majority of those who were forced to leave their homes because of the earthquake sought accommodations with friends and relatives, and about 20 percent used officially-designated congregate-care facilities (Bolin and

Stanford, 1990). Many victims camped outside their homes, a pattern that has been observed in other earthquake disasters in the U.S. and abroad, notably the 1983 Coalinga, 1985 Mexico City, and 1987 Whittier earthquakes (Tierney, 1985b, 1988).

The need for post-earthquake temporary shelter was linked to pre-earthquake housing problems in the affected region. Those most likely to need emergency shelter had low incomes and were renters who had lived in older dwellings that were in poor repair. Lower-income people were also likely to remain in public shelters longer because of their inability to find affordable housing after the earthquake.

Homelessness was a significant problem in the impact region before the earthquake that affected the post-impact provision of housing as well. Moreover, because homeless shelters and single-room-occupancy hotels tended to be older buildings with little earthquake resistance, those structures sustained high levels of damage, worsening the situation for those who had been homeless or at risk of homelessness before the earthquake. Conflicts developed after the earthquake because of relief agencies' efforts to distinguish between the "pre-disaster homeless"—whose hous-

Alexandria City, Missouri — A woman reflects in a shelter during the Midwest Floods of 1993.

ing problems, in their view, could not be attributed to the earthquake—and those with "legitimate" disaster-related housing needs. In cities such as San Francisco and Santa Cruz, some service providers tried to ensure that those who lost their homes to the disaster—"legitimate" disaster victims—would not have to mix with homeless and transient people in temporary shelters. Community and homeless advocacy groups contended that programs should try to meet the housing needs of everyone affected by the earthquake, including those who had been homeless before, rather than restricting eligibility to those who lost their homes in the disaster (Bolin and Stanford, 1990; Phillips, 1998).

Latino community residents in the Santa Cruz County city of Watsonville were particularly hard-hit by the earthquake, since they had a greater tendency than other residents to live in overcrowded conditions and in substandard housing that sustained extensive damage. Concerned that their needs would not be met in Red Cross-operated emergency shelters, afraid of seeking shelter in officially-designated indoor shelters because of aftershocks, and also fearful that if they did so agencies like the U.S. Immigration and Naturalization Service might question their right to be in this country, a group of Latino residents devised an informal and more culturally acceptable temporary sheltering arrangement in a city park. The presence of the unofficial shelter became a source of conflict between the community and official providers of disaster services. Because it was highly-publicized and supported by many community residents as well as by a number of outside organizations, the unofficial shelter also became a focus for mobilization within the Latino community and a mechanism through which those groups could press for other kinds of help following the earthquake. Rather than being a completely new pattern of action for this segment of the population, the organized protest against the sheltering arrangements that were available and the emergence of the alternative tent city were related to earlier political and labor struggles in that community (Simile, 1995).

One major lesson from recent studies on housing following disasters is that, as the major metropolitan areas of the U.S. become increasingly ethnically and racially diverse due to immigration and other population trends, the population requiring post-disaster sheltering and other services will reflect that diversity (Bolin and Stanford, 1990; Phillips, 1991, 1998; Bolin, 1993, 1994). One implication of such findings is that organizations providing shelter and housing for disaster victims must become more aware of and responsive to the needs of the different groups they will be serving.

Unfortunately, studies also suggest that, in attempting to arrange for the provision of post-disaster housing, organizations have a tendency to react to conflicts with disaster victims by defining the victims themselves as the problem, when in fact it may be their own activities that are the source of difficulty. Instead, the problem can more usefully be framed as a conflict between two very different cultures. The culture of the aid-giver is defined by the rule-bound requirements of administering a bureaucracy, while the culture of many aid-receivers is defined by the demands of living on the social and economic margins of society. Housing-related problems are often exacerbated by ineffective organizational mobilization, failure to take advantage of existing community resources, lack of interorganizational coordination, failure to recognize pre-impact conflicts and differences in community power, and poor intergroup communication (Quarantelli, 1982b).

Considering the importance of these issues, it is surprising that there has been so little research on disaster housing. In an overview of research in this area Quarantelli (1982b) identified a need for additional research on a range of topics. Some of those suggestions focus on organizational issues. For example, questions exist regarding interorganizational preparedness for post-disaster shelter and housing, community-wide coordination of housing operations, and local officials' understanding of disaster-related housing programs. Other issues singled out by Quarantelli as warranting further study center on the needs of victims and the psychosocial impacts of different types of post-disaster housing—for example, allowing victimized households to remain at their pre-disaster home sites versus relocating them to mobile home parks. Additional research questions concern the sheltering needs of specific populations, such as those who reside in institutions. New knowledge might be gained also through systematic cross-national research on sheltering and housing and on lessons that can be learned by studying refugee camps in developing nations. Studies are needed on the full cycle of post-disaster sheltering and housing, from evacuation and emergency shelter through permanent housing, and on optimal ways of providing for the housing needs of victims under different impact conditions. Research designed to obtain victims' own assessments of the housing process could prove particularly informative in revealing deficiencies in current policies. Unfortunately, in the two decades since Quarantelli's review, very little research of this type has been undertaken, and the majority of these issues remain to be addressed.

OTHER INDIVIDUAL AND GROUP BEHAVIOR DURING AND IMMEDIATELY FOLLOWING DISASTER IMPACT

Questions Regarding Panic and the "Disaster Syndrome"

For some types of disaster agents, such as riverine floods and hurricanes, forewarning is possible, and measures to protect the public concentrate on encouraging evacuation and other actions that are designed to protect life, safety, and property. However, other types of disasters, such as earthquakes, explosions, and many tornadoes, strike virtually without warning, and in these cases people must take action very rapidly in order to remain safe. Researchers have had a longstanding interest in understanding how people react during and immediately following disaster impact, particularly in events that strike with little or no warning. Authorities have also expressed concern about whether they should expect community residents to respond in an orderly and rational fashion when disaster strikes, or whether panic and other maladaptive behaviors will be common.

Quarantelli's pioneering study on panic (1954) was among the first empirical studies to explore in detail the kinds of behavior individuals engage in at the time of disaster impact. This work established that panic flight is extremely rare at any time—before, during, or immediately after disaster impact—and it identified the conditions under which panic does occur. Specifically, Quarantelli (1977) found that panic is more likely to occur when:

- There are pre-existing beliefs in a group that certain kinds of situations will lead to panic;
- Ineffective crisis management leaves people feeling completely on their own;
- People begin to feel that there is an immediate threat of entrapment. Panic does not develop when people know they are trapped, but rather when they sense that their chances for escaping danger are dwindling;
- People begin to believe that there is no possibility of saving themselves, except through flight; and
- People have a sense of complete social isolation—that is, that there is no one else in the setting upon whom they can depend.

The fact that these conditions are present in only a vanishingly small number of emergencies accounts for why panic is so rare. The literature on U.S. disasters consistently shows that social solidarity remains strong during the emergency response phase in even the most trying of circumstances, and few situations occur that can completely break down social bonds and eliminate the feeling of responsibility people feel for one another, especially for others whose lives may be in danger. As we note elsewhere in this volume, the notion that disasters engender pro-social, altruistic, and adaptive responses rather than negative reactions like panic flight continues to be among the most robust findings in the literature.

Continuing with this general research tradition, more recent studies—some employing systematic survey research techniques—have attempted to document in detail the actions people take during and immediately after disasters strike. For example, Bourque and her associates studied the behavior of community residents during the time of impact and in the immediate post-impact period in the 1987 Whittier Narrows, 1989 Loma Prieta, and 1994 Northridge earthquakes. Telephone surveys were conducted with systematically-selected samples of residents in high-impact areas to explore such topics as what people did during the actual period of earthquake shaking, the extent to which they engaged in self-protective actions, how fearful they were at the time of the earthquake, what use they made of the mass media immediately after impact, and whether and why they evacuated.

Their research found once again that people generally behave in an active and adaptive fashion during and after disaster impact. Many of those surveyed reported that they had taken the kinds of self-protective actions that authorities had recommended. In addition to adding further support to studies that find maladaptive behavior to be highly rare, these findings also suggest that earthquake awareness and preparedness campaigns have had some influence on people's behavior. However, there were still some people who engaged in actions that such campaigns had tried to discourage, such as running outside during earthquake shaking. Behavior during and immediately after earthquakes was influenced by a number of socioeconomic, situational, and social-psychological factors, including education, income, presence of children in the home, the person's location at the time of impact, and levels of fear residents experienced during earthquake shaking. (For detailed discussions of findings from the Whittier and Loma Prieta earthquakes see Goltz, Russell, and Bourque, 1992; Bourque, Russell, and Goltz, 1993).

Other studies have focused on behavior in other kinds of disasters, such as fires and explosions. Johnson (1988; Johnson, Feinberg, and Johnston, 1994), who conducted research on the 1977 Beverly Hills Supper Club fire in northern Kentucky, which killed over 160 people, found that social ties and a sense of responsibility for others persisted within the crowd that was trying to escape from the fire. Despite conditions of extreme peril, patrons inside the club exited in an orderly fashion, generally with their dining companions, and altruistic responses were far more common than competitive behavior.

These kinds of findings, which have also been documented for other situations involving fire (Canter, 1980; Keating, Loftus, and Manber, 1983), stand in stark contrast to the myth that panic and social breakdown are common in high-threat situations. The literature on fires contains many accounts that highlight the resourcefulness of fire victims and the extent to which, rather than panicking, they provide support to one another—even risking their lives to do so.

Aguirre, Wenger, and Vigo (1998) studied the emergency evacuation that took place in an equally perilous situation, the World Trade Center bombing of 1993. Their research focused on building occupants' threat perceptions immediately after the explosion, as well as on how they interacted to develop emergent definitions of the situations and emergent norms indicating appropriate lines of action. Once again, there were no documented instances of panic behavior; evacuation behavior was orderly, cooperative, and influenced by pre-emergency social ties.

While some images of disaster-related behavior focus on extreme behavioral reactions such as panic flight, others depict disaster victims as helpless and unable to act. Various reports have noted the presence after some disasters of a cluster of symptoms collectively called the "disaster syndrome." The syndrome is generally described as a state of shock characterized by docility, disoriented thinking, and a general insensitivity to cues from the immediate environment. One early discussion of this symptom cluster in the social science literature on disasters appears in Wallace's work on "mazeway disintegration" (1957). Wallace, an anthropologist, described the shock behavior that characterized surviving victims whose friends and family members had died. The behaviors closely followed what has been subsequently described in the psychiatric literature as "grief reactions" (Perry and Lindell, 1978). Over the years, three important conclusions have been drawn from the body of research on shock and passivity reactions in disaster situations. First, the disaster syndrome or shock reaction appears most frequently in sudden-onset, low- (or no-) forewarning events involving widespread physical destruction, traumatic injuries, and death (Fritz and Marks, 1954). Second, when shock reactions do appear, only a relatively small proportion of the total victim population is affected. Finally, the disaster syndrome is transient. In the very rare cases when it does occur, it usually persists for only a few hours and rarely lasts beyond the immediate post-impact period. In one of the few methodologically sound studies of the phenomenon, Fritz and Marks (1954) reported in a disaster report for the National Opinion Research Center that only 14 percent of their random sample of victims showed any evidence of the kinds of symptoms usually associated with the disaster syndrome.

In short, even though media reports commonly describe disaster victims as dazed, stunned, and disoriented, true cases of psychological paralysis in the face of disaster are extremely rare. However, this is not the same as saying that residents of disaster-stricken areas experience no short- or long-term emotional effects. As Singer has noted (1982: 248):

> Reports of actual experiences reveal that most persons respond in an adaptive, responsible manner. Those who show manifestly inappropriate responses tend to be in a distinct minority. At the same time, most people do show some signs of emotional disturbance as an immediate response to disaster, and these tend to appear in characteristic phases or stages.

Both victims and emergency response personnel do experience distress that can manifest itself in a variety of physical and emotional symptoms. Documented disaster stress reactions include sleep disruption, anxiety, nausea, vomiting, bed wetting, and irritability (Houts, Cleary, and Hu, 1988). Although most such reactions appear to be transient, some may become long term. For the most part, however, individuals affected by disasters appear to be able to develop coping mechanisms for these kinds of problems with little or no assistance from outsiders. For the majority, whatever short-term stress reactions do develop do not appear to interfere with the ability to act responsibly or to follow instructions from emergency officials. As with panic, isolated cases of debilitating shock have been documented in some disaster events, but they remain very rare and should be considered atypical.

Socially Integrative Responses

In contrast to the negative, dysfunctional images of disaster behavior discussed above—panic, disorganization, and helplessness—empirical research suggests that behavior in disaster situations is adaptive and problem-focused. Rather than being dazed and in shock, residents of disaster-stricken areas are proactive and willing to assist one another. Pro-social rather than anti-social behavior is the norm. As our later discussions on disaster volunteers and search and rescue activities show, key response tasks typically are performed by community residents themselves. People behave altruistically, often showing more of a sense of care and responsibility for fellow community residents than they would normally display during non-disaster times. Of course, this does not mean that emergency response behavior meets all the classic economic standards of rationality, but only that people attempt to make good choices, given the limited alternatives and the uncertainties they face.

Behavior in emergency situations is strongly influenced by pre-emergency behavioral patterns. What we see, in other words, is continuity between pre- and post-disaster behavior rather than discontinuity. One major exception to this pattern involves anti-social behavior, which tends to decline during the post-disaster emergency response phase. Specifically, looting is very rare following U.S. disasters, crime rates tend to decline following large-scale events, and, despite what many people believe, it has never been necessary to declare martial law following any U.S. disaster (Quarantelli and Dynes, 1972; Taylor, 1977; James and Wenger, 1980; Lindell and Perry, 1992).

The literature shows consistently that, at least in the post-impact response period, disasters promote cohesion among victims, as well as between victims and other community residents (Fritz, 1961b). Disaster victims themselves are generally the ones that initiate search and rescue and the provision of first-aid. Nonvictims in the impact region typically engage in helping behaviors directed at victims, and nonvictims from an even wider area typically help by donating materials. Indeed, Barton (1969) used the term "mass assault" to characterize the period immediately following impact in part because so many community residents become involved in helping fellow disaster victims during that time.

Along these same lines, Wenger (1972) has documented declines in some types of exclusionary social participation (in clubs, for example), reductions in the purchase of luxury goods, declines in the need for formal social control (e.g., for traffic offenses), and increases in mutual support activities. These findings reflect the development of what has been referred to as the "altruistic" or "therapeutic" community (Wilmer, 1958; Fritz, 1961a, 1961b; Barton, 1969).

Research suggests that the therapeutic community response is related to convergence behavior, another common pattern in disasters (Fritz and Mathewson, 1957; Boileau et al., 1979; Kartez and Lindell, 1987, 1990). During the immediate emergency response period, a stricken community often becomes the focus for aid-giving efforts by non-victims, those from surrounding communities, private organizations, and larger political units, such as counties, states, and the federal government. The result is an influx of equipment, goods, and people—some individuals acting on their own and some representing organizations. Often such aid arrives unannounced and in extremely large quantities.

Disasters are also accompanied by a more general sympathetic response on the part of nonvictims that is similar to the convergence response. Sympathetic behaviors tend to take the form of offers of direct help to victims, including food, clothing, and lodging. The earliest social scientific documentation of this type of response is found in Prince's study of the Halifax, Nova Scotia, explosion, in which he observes that (1920: 137):

> The idea spread of taking the refugees into such private homes as had fared less badly. It became the thing to do. The thing to do is social pressure. It may be unwilled and unintended but it is inexorable. It worked effectively upon all who had an unused room.

Since the time of Prince's study, considerable literature has developed that documents increases in helping behavior among both victims and nonvictims following disasters (Vallance and D'Augelli, 1982; Watson and Collins, 1982; Young, Giles, and Plantz, 1982; Aguirre et al., 1995; Beggs, Haines, and Hurlbert, 1996). Particularly in disasters occurring in Western societies (although the cross-cultural literature is growing), these kinds of altruistic and sympathetic behaviors typically become normative. The research reviewed here should not be interpreted as showing that disasters are always socially integrative. As discussed below, conflict can also occur during the period following disaster impact, and the research literature also documents departures from altruism in some disasters. But the evidence is extremely strong that the emotions and behavioral patterns that prevail in disaster situations are altruistic and positive. One role of future research will be to identify the conditions under which patterns fail to develop or break down.

Disaster Volunteers

The emergency response period is also marked by the involvement of large numbers of volunteers in activities aimed at coping with disaster-related problems. Research findings indicate that volunteer activity increases at the time of disaster impact and remains widespread during the emergency period, particularly in highly damaging and disruptive disas-

People volunteer on a massive scale during the 1993 Midwest Floods.

ter events. For example, in a survey conducted on a random sample of nearly 3,000 Mexico City residents following the 1985 earthquake, 9.8 percent of all those surveyed reported engaging in some sort of volunteer action during the three-week period after the earthquake. Extrapolated to the population of the city, this translates into at least 2,000,000 disaster volunteers (Dynes, Quarantelli, and Wenger, 1990). Following the 1989 Loma Prieta earthquake in California, a survey conducted in San Francisco and Santa Cruz Counties by O'Brien and Mileti (1992) found that a large majority of residents—70 percent in Santa Cruz and 60 percent in San Francisco County—participated in some type of emergency response activity following the earthquake, including helping with search and rescue activities, providing food and water, assisting with clean-up and debris removal, and providing shelter to displaced earthquake victims.

The examples above focus on volunteer activity that emerges spontaneously after disaster impact. However, volunteer behavior can take other forms. In some cases, volunteering is more or less institutionalized in disaster situations. Probably the best-known example of institutionalized volunteering is the Red Cross, which plans extensively to recruit and mobilize volunteers to meet disaster-related needs. Britton, Moran, and

Correy (1994) have documented the activities of "permanent emergency volunteers," who regularly get involved in response activities. Many groups whose involvement in the response and recovery phases following disaster has become routine (e.g., emergency radio communications groups and ecumenical groups such as the Church of the Brethren and Mennonite Disaster Services), are made up wholly or partly of volunteers. Additionally, extensive volunteer behavior takes place within existing organizations in disaster situations (Stallings, 1989).

Despite its importance, volunteer behavior has not been studied extensively in the disaster research field. We know relatively little about spontaneous volunteers and even less about the other patterns of volunteer behavior described above. Not much is understood about which social groups volunteer and why. Wenger and James (1994) note, for example, that while earlier research found that men and young people are particularly likely to act as disaster volunteers, other studies find that women volunteer more frequently and that young people are not disproportionately involved. Their own research on the Mexico City earthquake indicated that males were more likely to engage in search and rescue, while females were involved in the provision of food and supplies. This suggests that post-disaster volunteering is influenced by pre-disaster roles—another example of continuity between pre- and post-disaster behavior patterns. Additionally, adults between 18 and 44 years of age and lower-status residents were found to have been more likely than other groups to become involved in search and rescue, although why this was the case is unclear.

These variations in research findings suggest the need for further investigation of volunteerism as a social phenomenon during disasters and for the development of typologies of organized volunteer behavior. One potential line of research would be to examine the involvement of volunteers in each of the emergency response activities identified in Chapter Two, Table 2.3. A cursory review of the literature suggests that volunteers have been involved in all four response areas—emergency assessment, expedient hazard mitigation, population protection, and incident management. Research should also address the factors that influence both patterns of volunteer behavior and the effectiveness of volunteer efforts. Those factors are likely to include the characteristics of disaster events, such as their severity and scope of impact and the nature of the damage and disruption they cause; characteristics of affected communities, such as prior disaster experience and the existence of disaster subcultures; and characteristics of affected populations, such as the extensiveness of their social networks and their involvement in civic activities during non-disaster times.

Emergent Groups

Research on the phenomenon of group emergence overlaps to some degree with work on disaster volunteers. Earlier work on organizational response activities highlighted the changes organizations undergo as they adapt to handle crisis-related demands. What has come to be termed "the DRC typology" (because it was originally developed at the Disaster Research Center) has been used extensively to characterize how organizations adapt (see Dynes, 1970; Brouillette and Quarantelli, 1971; Stallings, 1978). The typology classifies responding organizations along two dimensions—tasks and structure—and according to whether or not either dimension undergoes change during the emergency period. This classification yields four types of organizational responses to disaster (see Figure 3.3). Established, or Type I, organizations perform the same tasks during disasters that they usually carry out when disasters do not occur. Expanding, or Type II, organizations tend to be small or relatively inactive during non-disaster periods, but they increase in size or undergo changes in structure during the emergency while performing tasks similar to the ones for which they are normally responsible. Extending, or Type III, organizations retain their pre-disaster structure but engage in disaster-related tasks that are new. What changes for these organizations is what they are doing in the emergency situation, not their membership

Tasks

Organizational Structure		Routine	Nonroutine
	Same as Predisaster	Type I Established	Type III Extending
	New	Type II Expanding	Type IV Emergent

Figure 3.3 Typology of Organizational Adaptation in Crisis (Based on illustration in Dynes, 1970)

or authority structure. Emergent, or Type IV, organizations are newly-formed entities that were not part of the pre-disaster community setting; such groups typically are informal and relatively undifferentiated structurally, consisting mainly of residents of the stricken area, at least initially.

Changes in organizational structure and functioning and group emergence invariably accompany major disaster events. The degree to which a disaster requires extensive organizational adaptation and stimulates the emergence of new groups may be one measure of disaster severity and may also provide some indication of which disaster events will be particularly difficult to manage. What emerge in disaster situations are new behavioral expectations (norms) and social structures that "represent populations of systems being born" (Drabek, 1986: 267). Often the persons who participate in emergent groups have little or no experience performing disaster-related tasks; their relationships and roles are untried and new. For these reasons, groups may have difficulty getting themselves organized. At the same time, such emergence can greatly benefit affected communities, since disasters strain existing resources. Local residents are often the best judges of what they need, and they bring a detailed knowledge of the disaster-stricken area that enhances their ability to respond effectively. Thus, local disaster "victims," rather than behaving helplessly and waiting for outside aid, actually play a vital role in the post-disaster response.

Research on post-disaster search and rescue (SAR) illustrates this point. The effective search of damaged structures and the rescue of victims who are trapped are critical tasks in the emergency response period. Studies concur that emergent groups play a major role in SAR, particularly in the initial period following disaster impact. After the 1992 gas explosion in Guadalajara, Mexico, for example, Aguirre and his colleagues (Aguirre et al., 1995) interviewed 43 victims who had been buried alive in the impact area, as well as local SAR volunteers. None of the victims had been trapped in the rubble for more than two hours, and all had been rescued by relatives, neighbors, and others who lived in the immediate area affected by the explosion. Professional SAR resources arrived at the scene too late to have much of an impact on victim survival; the vast majority of the victims they located and extricated were already dead. This pattern of extensive and effective involvement by emergent groups in SAR activities has been documented for a variety of disaster agents and in different cultural settings (see, for example, Noji, 1989, on the 1988 Armenian earthquake; Dynes, Quarantelli, and

Wenger, 1990, on the 1985 Mexico City earthquake; and Tierney, 1994, on the 1989 Loma Prieta earthquake; for an overview of earlier research, see Wenger, 1991).

Early work by Dynes and Quarantelli (Quarantelli, 1966; Dynes and Quarantelli, 1968; Dynes, 1970) stressed the ubiquity of emergence during disasters. An initial study by Parr (1970) identified emergence as a common phenomenon that occurs most typically in situations characterized by a lack of pre-planning, ambiguity over legitimate sources of authority, authority structure collapse, perceived inadequacies in organizational performance, and exceptionally challenging or newly-generated disaster tasks. Put simply, emergence occurs because of a "sharp increase in demands . . . accompanied by a high degree of organizational impairment" (Parr, 1970: 4).

Another earlier study (Gillespie, Mileti, and Perry, 1976; Mileti, Gillespie, and Perry, 1975; Perry, Gillespie, and Mileti, 1974) followed the emergence and subsequent formalization of a disaster-connected emergent group. These authors found that a small group that arose in response to what had been initially considered a transient disaster-generated need for temporary shelter went on to develop an ideology, formal cadre, and organizational structure much like a growing social movement. In the aftermath of the disaster, the group transformed its goals to address more general community needs and persisted as an organization and a force in community politics for nearly a decade. The authors concluded that the organization was able to persist because the disaster-related need for shelter was related to a larger need for welfare services in the community.

Since the time of the first assessment, several other studies have focused on factors related to emergence as well as on the consequences of emergent group activity. A key study conducted by the Disaster Research Center in the early 1980s focused specifically on "emergent citizen groups" (ECGs) (Quarantelli et al., 1983; Quarantelli, 1985; Stallings and Quarantelli, 1985). The ECG project looked at approximately 50 pre- and post-disaster emergent groups from all regions of the United States and in a variety of disasters, including floods in Kentucky, hurricanes in Texas, and landslides in California. Situations involving chronic hazards such as radioactive waste and air pollution were also studied.

The study found that ECGs are typically composed of a small active core who participate for the entire time the group is active, a larger supporting circle, and a still larger number of nominal supporters—in essence three tiers of participation. ECGs develop both before disasters (to prepare) and after disasters (to respond). Preparedness ECGs tend to be community-oriented and driven by broad-based concerns, such as the threat of a nuclear power plant accident. Response ECGs are more task-oriented and are more likely to form after very severe disasters. ECGs usually have few monetary resources, but funds are not essential to success. Rather, having volunteer time and commitment are most essential for mobilization.

Although much has been learned about emergence, the topic remains understudied. In fact, 15 years ago, Drabek (1986) called for a theory of emergent structures that has yet to appear. He suggested that several interrelated issues need to be considered. First, what facilitates emergence? What factors shape emergent systems? Do major contributors include structural strain, the idea that something can or should be done, or ambiguity over authority? Second, what structure does the emergent organization assume, and what factors shape those emergent structural properties? Third, what accounts for stability and variation in emergent group phenomena? Fourth, when does emergence end and why? (For additional material on emergence, see the discussion below of Drabek's research on emergent multi-organizational networks.)

Based on his comprehensive review of the literature, Drabek (1986) argued that emergent structures originate when: (1) there is organizational atomization and a lack of overall community coordination during the emergency period; (2) there is ambiguity over authority; (3) people are isolated from emergency organizations and information; and (4) prior disaster experience is minimal. Drabek's own work on emergent social networks, discussed below, identifies a number of factors that also influence emergence, including event qualities, demands, community emergency response capability, pre-event communications patterns, domain consensus, interpersonal linkages, and resources.

Recent research points to various conditions that are likely to foster emergence. Generally speaking, those conditions include a legitimizing social setting, a perceived threat, a supportive social climate, pre-existing social ties, and the availability of resources (Quarantelli et al., 1983). In the pre-disaster setting, the more a collectivity of concerned persons comes to define a condition as posing a threat, the more likely it is that an ECG will appear. The presence of a visible target, such as a landfill generating noise and unsightly debris or a chemical facility producing noxious smells, may make ECG development more likely. Repeated exposure to highly-damaging events such as hurricanes or earthquakes may also prompt the emergence of ECGs.

Research also suggests that social and political inequality are additional factors driving emergence. Dominant groups that produce plans and preparedness measures based on their own cultural norms, values, and expectations may fail to address the needs of minority groups within the community (Neal and Phillips, 1995). When this happens, neglected groups may organize themselves, either to provide their own assistance to group members or to press for assistance from official service providers. Similarly, inflexible bureaucratic structures and procedures may result in a failure to meet victims' needs, leading to conflict and group emergence. Emergent activity may thus be one avenue through which previously marginalized populations obtain the help they need. For example, after the Loma Prieta earthquake Latino citizens who concluded that they were not receiving adequate aid or information in Watsonville came together to protest their exclusion from earthquake relief activities (Phillips, 1993; Simile, 1995). As a result of these protests, the City of Watsonville appointed a Latino ombudsperson to work with the city throughout the recovery, rewrote its disaster plan to make it more of a community-based plan, and hired a bilingual emergency manager.

In summary, research conducted since the first assessment has lent further support to earlier studies that found that most of those who are affected by disasters respond constructively. Heightened social solidarity, prosocial behavior, and intensive community involvement in response activities are patterns that have been documented for decades in events ranging from natural disasters to human-caused tragedies such as the bombing of the Oklahoma City Federal Building. Following impact, ambulatory victims routinely search for survivors, care for victims as resources permit, and protect others from further harm. It is important to emphasize these kinds of research findings because, besides being erroneous, misconceptions about irrational and antisocial behavior in disasters can also hamper the effectiveness of emergency planning and response by leading authorities to misallocate resources and misinform the public.

ISSUES FOR FUTURE RESEARCH

Throughout this chapter we have made the point that, while social behavior during the emergency period has been a major focus for study since the field of disaster research began 50 years ago, a number of gaps remain in the literature. In the area of household protective responses and evacuation—probably the most-thoroughly researched of all response-related topics—most studies have focused on single cases, and little comparative research exists. It is impossible to disaggregate community-level, household-level, individual-level, and agent-specific influences on protective responses without more systematic comparative research. An overarching need in the warning/evacuation area is to develop a common approach to operationalizing and measuring concepts so that research results are comparable and cumulative. Without more comparative studies using consistent measures, we will be unable to say with confidence how effective different types of warning systems will be with different populations and in different disaster situations. More broadly, we still know far too little about the psychological, social, economic, and political factors that influence the public's response to warning messages.

The same can be said for the other topics we considered in this chapter. Many aspects of emergency sheltering and short-term housing are not well understood. In particular, there is a need to learn more about how to facilitate the transition from the sheltering and housing arrangements people make (or that are provided for them) in the immediate aftermath of disasters to more permanent housing. As noted above, existing research suggests that a number of factors influence that process, including household resources, the availability of insurance coverage, both disaster-related and other governmental housing assistance policies, the manner in which housing-related services are provided to victims, and the availability of housing alternatives for households at different income levels.

With respect to group behavior during the impact and emergency periods, we have also identified a number of important topics on which research is seriously lacking. We still understand little about why emergent groups form, what facilitates emergence, how emergent structures develop, and why some emergent groups persist while others disappear. In Japan, 1995 became known as "the first year of the volunteer" because individuals and groups had volunteered in unprecedented numbers at the time of the Kobe earthquake. The fact that a disaster-related pattern that U.S. research takes almost for granted—the convergence of volunteers and the formation of emergent groups—was considered remarkable in Japanese society shows how much societies can differ, as well as how much still remains to be learned about group behavior in disaster situations. Relatedly, recent research on the important role played by community-based organizations in responding to disasters in this society (Bolin, 1998) points to a need to better understand how grassroots organizations that originally formed to meet entirely different community needs become involved in the provision of disaster-related services. Although the important contributions made by emergent groups and community residents during the disaster response period have long been acknowledged, broader theoretical questions about the relationship between governmental organizations and civil society institutions, both in the U.S. and in other societies, remain to be addressed.

In closing, it is also important to emphasize that much of what we claim to know about the public response to disasters is based on research on the white majority population. With so little research on minority residents' responses in disasters, what degree of confidence can researchers have in their conclusions and policy recommendations? Researchers have only recently begun to address racial, ethnic, and social class influences on disaster-related behavior, and the little work that has been done highlights the need for further, in-depth research.

CHAPTER FOUR

Meeting the Challenge: Organizational and Governmental Response in Disasters

O RGANIZATIONS RESPONDING DURING disaster situations face a number of challenges. Upon notification of an actual or imminent disaster they must mobilize, assess the nature of the emergency, prioritize goals, tactics, and resources, and coordinate with other organizations and the public, while making an effort to overcome the operational impediments posed by the disaster (Kreps, 1991). All of these activities must be accomplished under conditions of uncertainty, urgency, limited control, and limited access to information. In the absence of prior interorganizational and community planning, each responding agency will tend to perform its disaster-related tasks in an autonomous, uncoordinated fashion (Kartez and Lindell, 1987). Indeed, one of the challenges of disaster planning and management is to overcome the natural tendency of organizations to maintain their independence and autonomy and to encourage them to have a broader interorganizational and community-wide focus.

Because of the crucial role organizations play in emergencies, organizational response has been a key focus of disaster research since the field began. Initial efforts at systematizing research findings on organizations coincided roughly with the activities of the first research assessment nearly three decades ago. Dynes's *Organized Behavior in Disaster* (1970) was among the first published works that attempted to synthesize findings on organizational response during disaster situations. A 1970 special issue of the *American Behavioral Scientist* (Quarantelli and Dynes, 1970b) contained a series of articles focusing on the disaster-related activities of various organizations, such as departments of public works, hospitals, and fire and police departments, as well as private agencies such as the Red Cross and Salvation Army. *Human Systems in Extreme Environments* (Mileti, Drabek, and Haas, 1975), written in conjunction with the first assessment, contained a series of research-based propositions related to organizational adaptation and response during high-stress situations. Included in that volume were discussions of the four-fold typology of organizational adaptation introduced in Chapter Three of this volume and the demand-capability model outlined in Chapter One. Based on a review of existing literature conducted in the mid-1970s, Dynes and Quarantelli (1977b) developed an inventory of propositions related to organizational communication and decision making in crises. These works, which were qualitative and based primarily although not exclusively on case-study material, formed the core of what was known about organizational response a generation ago. (Drabek's laboratory simulation work [Drabek, 1965; see also Drabek and Haas, 1969] is a conspicuous exception to this pattern.)

Early work on organizational response was primarily inductive. Researchers analyzed descriptive material (such as interview transcripts or case studies based on specific disasters) and developed empirical generalizations based on those examinations. Such research methods have not changed appreciably over time even though social science research in general has become more quantitative. Research on organizational response contrasts significantly with studies on household protective response activities and on household and organizational emergency preparedness, both of which increasingly use standardized questionnaires, survey sampling methods, and hypothesis-testing approaches.

The fact that emergency response research has remained largely qualitative and case-study oriented is undoubtedly due in part to the interests, training, and methodological preferences of scholars who conduct research on the way organizations behave during disasters. However, it is also a consequence of the comparatively modest levels of funding that have been available for this type of research and the practical difficulties associated with carrying out large-scale, quantitatively-oriented organizational response studies.

The most common way organizations have been studied in the disaster literature is as participants in community-wide response networks. Organizations with officially-designated disaster roles (e.g., police, fire, emergency medical services, emergency management) are almost always involved in major disaster events. An organization's involvement and centrality during the disaster response period is determined both by official designation and by the extent to which it possesses emergency-relevant resources such as trained personnel, equipment, information, and facilities. Organizations that do not have clear-cut disaster missions may also become involved in response activities when they have needed resources. For example, builders and contractors may act as extending organizations in a response network by lending their personnel and equipment for search and rescue or debris removal. Following the Kobe earthquake, neighborhood schools provided shelter, which was an intended function, but also served as improvised medical care facilities, community information centers, and temporary morgues, because they were often the only community institutions to which victims could turn. The overwhelming need for emergency shelter also led owners of many private businesses and office buildings to open their doors to victims needing shelter.

Beyond their role in community-wide emergency activities, organizations can also become responders either because they are directly affected by the occurrence of a disaster or because they are directly responsible for causing one. Like households, organizations can be "victims" of disaster, required to respond, to meet the needs of their workers and customers, and to recover in emergency situations. We currently know less about these aspects of organizational involvement during disasters than about their role as community responders.

Following the same strategy we used in our review of preparedness research in Chapter Two, we will begin by discussing emergency response activities undertaken by organizations whose domains most clearly include disaster-related responsibilities, and then move to consider research on how other organizations respond in disaster situations.

RESPONSE ACTIVITIES OF EMERGENCY MANAGEMENT AND OTHER CRISIS-RELEVANT ORGANIZATIONS

Local Emergency Management Agencies and Emergency Operations Centers

Several studies conducted in the 1970s and 1980s focused on the preparedness activities undertaken by emergency management agencies at the local level (Dynes and Quarantelli, 1977a; Hoetmer, 1983; Caplow, Bahr, and Chadwick, 1984; Drabek, 1985). Wenger, Quarantelli, and Dynes (1986: 8) summarized work that had previously been done on emergency management organizations this way:

> There was general agreement that variability in the overall functioning of local emergency management agencies exists within the United States. They are generally small organizations or offices. Generalists, rather than specialists, tend to prevail . . . [t]here is little hierarchial differentiation which results in easier internal communication and clearer notions of responsibility.

They also noted several ways in which local emergency management agencies showed variation: in their assigned responsibilities; in their relationships with other community organizations; in their methods of performing their emergency-related tasks; and in the quantity and kinds of crisis-relevant resources under their control.

Early emergency response research pointed to some of the difficulties that local emergency management agencies had in actually managing the response during disaster situations—a problem that has persisted. The review concluded that "although civil defense agencies often stated that the desired goal of their operation was the 'coordination of response,' in fact most of their activities did not involve management, or even coordination" (1986: 10–11). Instead, the agencies concerned themselves primarily with gathering and disseminating information and locating needed resources. At the time that overview was conducted 15 years ago, many local emergency management agencies lacked the capacity to actively manage and coordinate community-wide emergency activities. Underfunded and understaffed, lacking stature and authority in the multiorganizational disaster response network, and typically positioned organizationally at a distance from centers of power in local government, local civil defense organizations were not well-equipped to take over the management and direction of major emergencies.

Other studies, including more recent ones, suggest that this situation has improved over time, although those improvements have been uneven. A study conducted in the mid-1980s by the Disaster Research Center focused on the response of local emergency management agencies during six disaster situations. The purpose of that research was to assess the extensiveness and effectiveness of those agencies' response activities. Extensiveness was conceptualized as the degree of involvement emergency management agencies had in key disaster-related tasks such as evacuation, medical care, and sheltering. Effective operations were defined (Wenger, Quarantelli, and Dynes, 1986: 21) as those characterized by:

> excellent information collection and distribution, a fully-staffed and functioning EOC [emergency operations center] adequate human and material resources, a specialized division of labor among responding units with the coordination of those units by one agency, a legitimated authority structure, integrated and coordinated relationships with outside organizations, mutually beneficial and effective relationships between emergency officials and mass media representatives, and 'reality-based' activities.

Response extensiveness—that is, the number of required tasks in which emergency management agencies participated—ranged from limited to broad in the six events studied. Extensiveness was found to increase with prior disaster experience and the interorganizational breadth of pre-disaster planning. Similarly, effectiveness varied considerably across the six events. Emergency management agencies experienced the most problems in the areas of communication, the assignment and coordination of tasks, and authority relationships. Those problems tended to be interrelated.

Like extensiveness, effectiveness was related to previous disaster experience. Where pre-disaster planning was limited, responses also tended to be ineffective. However, while more extensive pre-disaster planning was associated to some degree with response effectiveness, that was not always the case. Ineffective responses were seen even in cases where planning was judged to be of high quality. Federal government involvement in supporting response planning was generally seen as having a positive impact on both preparedness activities and response effectiveness.

This particular study was based on too few cases to draw definitive conclusions, but it did make a number of important contributions. It identified different patterns of emergency management agency integration within local governmental structures, pointed out differences among comprehensive planning and response activities, and discussed the operational implications of these varying organizational patterns. Eight different patterns of organization were identified, ranging from those in which emergency management agencies were weak, isolated, or bypassed during the emergency response period to those that were well-institutionalized and embedded in communitywide emergency management systems.

One of the major changes identified in the Wenger, Quarantelli, and Dynes review (1986) was the increased use of EOCs in the management of emergency response operations. In a chapter directed to emergency management professionals, Perry (1991) discussed the functions of EOCs and identified several requirements that must be addressed in order for them to function effectively during disaster situations. Among these requirements are that procedures must be established for both the activation and the deactivation of the EOC and that the facility must be supplied with needed personnel and equipment. Moreover, management and communications systems must be adequate for the tasks they will be expected to perform during disaster situations, and care must be taken to ensure that those systems will remain functional following disaster impact.

In early work on EOCs and their effectiveness, Quarantelli (1978) indicated that EOCs can be effective in fostering interorganizational communication and coordination. However, he also discussed some of the problems that can accompany the use of EOCs in disasters. For example, EOCs themselves can receive damage and be forced to relocate. They can become overcrowded, the number of EOCs and command posts can proliferate, and questions can develop about who is actually in charge of EOC operations. They also can fail to function as intended.

Since the time of these earlier studies, very little research has focused on how EOCs actually perform during disasters or what makes for an effective EOC operation. An exception is a study by Scanlon (1994) which focused on 19 different disaster incidents, all of which occurred in Canada, in an effort to assess the extent to which Quarantelli's earlier findings were applicable to the situation in that country. EOCs were used in 13 of the 19 disasters Scanlon studied, and their organization and operations varied considerably across events. Sometimes EOCs were set up in special areas that had been pre-designated for that function; sometimes they were not. Some EOCs were set up according to plans; others were not. Participation by agencies and public officials in EOC activities varied. The police and fire department and the mayor's office almost always were represented, and agencies with hazardous materials expertise were often included if the incident warranted it. Beyond that, there was little consistency across events in which agencies were involved in EOC activities. Participation also fluctuated over time, with agency representatives coming and going at different phases in the response.

Like Quarantelli, Scanlon found that "[a]n EOC is an effective way to achieve coordination among agencies responding to a major emergency or disaster. The absence of an EOC seems to encourage the opposite" (1994: 70–71). At the same time, he found that the EOCs in the events he studied experienced many of the same kinds of difficulties the earlier research had identified, especially overcrowding and problems with being forced to relocate during an emergency. Perhaps the most important contribution of Scanlon's research is to suggest the range of different management and decision-making styles that were used in those events. We know of no comparable comparative research that has been conducted on EOC operations in recent U.S. disasters, and another look at how they function in actual events is long overdue.

As we discuss in more detail in Chapter Six, a number of important trends—including the information technology revolution and the marked increase in the professionalization of the emergency management field—have the potential for transforming emergency management in the United States. At this point, however, we lack detailed information on how emergency operations are currently managed at the local level in U.S. disasters, and so we lack a basis for evaluating the impact of the changes that are occurring. Most of the work that has been done has focused on specific disaster events and single communities rather than across events and communities. Further, comparative case studies of organizational emergency response have been limited in their ability to control for factors such as disaster event characteristics and pre-existing differences in community resources, which likely have a major impact on the performance of the emergency management function. Finally, different methodological approaches and concepts for assessing the performance of emergency management agencies have been used, and as a result, it is difficult to generalize from the research that has been conducted.

Fire, Police, and Emergency Medical Service Providers

We noted above that while fire and police departments are among the most important core organizations in the response system, relatively little is known about how they plan for disasters. Similarly, studies on organizational activities during the emergency response period seldom focus specifically on these organizations—unless, as sometimes occurs, the fire or police department happens to be the organization that serves as the local emergency management agency in a particular community.

In the only study on this topic of which we are aware, Wenger, Quarantelli, and Dynes (1989) looked specifically at fire and police departments in eight communities that were stricken by major disasters. The study focused on pre-disaster structure and planning, the disaster-related tasks performed by those organizations, and patterns of inter-

Seattle, Washington, March 6, 2001 — Federal, state, and local responders gather in the King County Emergency Management Department to discuss ongoing earthquake damage assessment activities.

organizational and intraorganizational adaptation. It also looked in some detail at the use of the incident command system (ICS), an organizational framework for managing emergencies that was developed in the fire service and that has achieved widespread acceptance in the emergency management field.

Several dozen empirical generalizations concerning variations in organizational structure, task performance, and response effectiveness were developed from this research. For example, structural alterations within police departments appeared to be more likely when a disaster was extensive, when resource levels were low, and when there had been little prior planning. Moreover, police department decision-making became more diffuse during disasters than in non-disaster times, and problems with communication and convergence were common. In contrast, fire departments underwent fewer organizational changes during disasters and generally had fewer operational problems than police agencies. Fire and police departments resembled one another in their preference for a high degree of autonomy and domain control in their everyday operations, and these patterns carried over into the emergency response.

As is the case with the other emergency-relevant organizations discussed here, there is only a small body of work on how providers of EMS perform in disasters. During the mid-to-late 1970s, the Disaster Research Center conducted research on the operations and effectiveness of EMS organizations and networks of service providers in 44 major disasters and mass-casualty situations. Quarantelli's *Delivery of Emergency Medical Services in Disasters* (1983) summarized that work and offered a number of generalizations about the functioning of EMS systems in disaster situations. One of his most important observations is that post-disaster search and rescue activities are typically performed by persons outside the formal EMS system, which reinforces the point made above about the extensive involvement of ordinary community residents in search and rescue. Quarantelli also found that the transportation of disaster victims to hospitals is almost invariably uncoordinated and that there is usually an oversupply of EMS resources, especially transportation resources like ambulances, following disaster impact. Additionally, triage, while sometimes attempted, tended to be "informal, sporadic, and partial" (1983: 76). EMS responses were characterized by a considerable degree of emergence, and central control of emergency care activities was rare.

In a related work, Tierney (1985a) traced the patterns observed in the delivery of medical services following disasters to broader factors that shape relations among health-care service providers on an every day basis. These factors include conflicts that exist between high- and low-status hospitals and between public and private service providers, professional hierarchies, and the high degree of jurisdictional complexity involved in providing emergency health-care services. Such problems do not disappear in disaster situations but rather are exacerbated by them, making it even more difficult for EMS organizations to operate in the uncertainty and urgency of the disaster environment.

A smaller follow-up to this earlier EMS project, which was conducted by DRC in the late 1980s and early 1990s, focused on the EMS response following eight natural and technological disasters (Tierney, 1993). That study found both continuities and discontinuities with earlier work. EMS resources were found to be adequate to handle the events studied, and the EMS systems in the eight communities had been involved in preparing for disasters prior to their occurrence. However, search and rescue and the transportation of victims to hospitals were generally ad hoc and uncoordinated. Moreover, trained, specialized search and rescue personnel played an important role only in emergencies with very localized impacts, such as plane crashes. Since so little of what was done to aid victims immediately after impact was under the control of official EMS providers, little formal triage was done in the field. Convergence was a major factor complicating the response in the events studied, and EMS responders experienced a range of difficulties with emergency communications. Response activities generally lacked central coordination, especially in disasters that had a wide geographical area of impact (Tierney, 1993). Thus, while change has occurred along some dimensions, older patterns evidently persist. Looking at post-disaster triage in light of earlier and more recent research, Auf der Heide has commented (1989: 11) that:

> . . . there are those who may believe that triage has improved since the Disaster Research Center studies . . . Although there may well have been improvements, some evidence suggests that many of the disaster response problems that were present in the '50s, '60s, and '70s are still seen in some form in the '80s.

And, we suspect, things have changed little since then.

A related field, the study of disaster epidemiology, deals less with the organization of emergency medical services for disaster victims than with the incidence and causes of disaster-related death, injury, and illness. In a recent volume entitled *The Public Health Consequences of Disasters* (1997), Eric Noji lists numerous knowledge gaps and research priorities related to the health impacts of different types of natural disasters. Key research needs that cut across different disaster agents include studies on such topics as the identification of risk factors for injury and post-disaster illness and the development of methods for protecting vulnerable populations from injury. Other research needs include the identification of behavioral factors related to disaster victimization and of optimal search and rescue strategies, as well as on the effectiveness of disaster warning systems for reducing mortality and morbidity. More generally, there is a need to develop standardized methodologies, measures, and data collection strategies so that comparable, cumulative data can be obtained.

As in many of the other areas we discuss here, the criteria for judging organizational and system performance in the provision of disaster EMS are not as clear as they need to be. We do know that requirements for evaluating and treating patients in disaster situations differ from those that characterize service delivery in routine emergencies, such as individual house fires and traffic accidents (Quarantelli, 1983). In major disasters, treatment procedures must be simplified, patient care may need to be rationed through triage, and it may be necessary to institute treatment procedures on a large scale in the field, rather than in medical facilities (Noji, 1997). Quarantelli (1983) has argued that effective provision of EMS during disasters must be based on the recognition that disasters are qualitatively different from "everyday" emergencies. However, little research exists directly linking planning assumptions to service delivery effectiveness.

PRIVATE-SECTOR ORGANIZATIONS IN DISASTERS

As noted above, private-sector organizations may be called upon to support community emergency response activities by providing equipment, facilities, and trained personnel. However, even if they are not involved in community-focused emergency response efforts, organizations affected by disasters must respond to disaster-related demands by providing for the safety of their employees, relocating their operations if their facilities are damaged, and using other coping strategies to minimize disruption to their own operations. Organizations may also find themselves responding to disasters of their own making, such as explosions, chemical releases, fires, and oil spills.

Until fairly recently, studies on the response of private-sector organizations during disasters were virtually nonexistent, and to date very few systematic studies have been done on the topic. A survey of research published in the last five years by the *International Journal of Mass Emergencies and Disasters* and the journal *Disasters* found only a handful of articles that focused specifically on private organizations. Even journals such as *Industrial Crisis Quarterly* and the *Journal of Contingencies and Crisis Management,* which might be expected to contain empirical research on disasters and businesses, have actually published very few empirically-based articles. Many studies that do deal with disasters involving private-sector organizations tend to focus not on the organization as a responding unit, but rather on how public-sector responders cope with the crises those organizations caused (see, for example, Harrald, Cohn, and Wallace, 1992, on the *Exxon Valdez* oil spill). We thus know very little about how businesses organizations actually respond when faced with disaster-related demands. Existing studies tend to focus on particular types of organizations and rather narrow topics. They also typically use small and non-representative samples, limiting the generalizability of their findings.

Drabek's (1994) study on disaster evacuation in the tourist industry focused on businesses in six communities that had actually been involved in emergency evacuations. Interviews with tourist-industry executives indicated that they engaged in many of the same kinds of behaviors and decision processes as community residents in deciding what to do about disaster warnings. For example, they tried to confirm warnings by consulting outside information sources. However, in seeking confirmation, they tended to turn to government officials for information to a much greater degree than residents. Not unexpectedly, they also turned to higher-up corporate executives and facility owners in making those decisions. Downplaying the danger was also a common pattern.

As evacuations proceeded, managers had to address a range of issues, such as deciding what information employees should give customers and then seeing to it that the information was conveyed; making alternative sheltering arrangements for evacuees, including both clients and employees; dealing with employee concerns, including the question about whether they would be paid during the evacuation period; providing transportation; arranging for security; and planning for the return of evacuees.

The protective responses undertaken by these organizations varied in extensiveness from not evacuating (i.e., keeping customers and employees at the site), to various forms of partial evacuation, to total evacuation, in which everyone was asked to leave and the facility was closed down. The extensiveness of evacuation activity was associated with characteristics of the individual managers, the evacuation messages that were given, and organizational and community-level factors. Specifically, managers who had viewed the probability of evacuation as likely were more likely to have evacuated, as were women, although the latter association was not strong. If the initial evacuation warning was received from some official source, such as the police or local elected officials, it was more likely to be heeded. Lodging establishments were more likely to evacuate than other tourist-oriented businesses, and size and prior experience with evacuation were positively associated with the decision to evacuate. The existence of a disaster subculture was also associated with evacuation, but weakly.

In a similar study, Vogt (1991) focused on a sample of 65 cases involving nursing home evacuations in an effort to determine the factors that were associated with effective evacuations. Slightly more than half of these evacuations were caused by weather-related events, but one-third were associated with mechanical failures, explosions, or chemical releases near the facility, as well as other causes. Evacuation effectiveness, measured in terms of the time taken to evacuate the facility, was found to be most strongly associated with three types of variables. First, effectiveness varied with the type of threat facilities experienced, with non-weather-related events eliciting a more rapid response. Population density also predicted effectiveness and was inversely related to evacuation time. Finally, the number of outside resources used in the evacuations also had an effect, with fewer sources of external aid making for speedier evacuation. This latter finding seems counterintuitive, but the author reasoned that evacuations defined as non-urgent allowed more time for outside help to mobilize, or alternatively that the mobilization and coordination of many different organizations was itself a factor that slowed down the evacuation process.

Surprisingly little is known about how private firms manage disasters, in part because, as we noted above, research has tended to focus mainly on the role played by those organizations in communitywide activities. When business organizations have been the focus of research, studies have tended to use small samples and to concentrate on particular types of organizations and rather narrow topics. To some extent, these sampling problems are a result of focusing on disaster events as units of analysis. For example, Three Mile Island, the Exxon oil spill, and the Bhopal disaster were very different disasters, even though private-sector organizations were central to all three. Systematic research on business organizations is also complicated by the fact that organizations differ greatly in terms of size, complexity, the technologies they use, and the extent to which they are required to plan for disasters.

Organizational Theory and Disaster Research

In recent years, interest in disasters has grown in the disciplines that study organizations, including organizational sociology and management-related fields. This research has focused primarily on private-sector organizations but also has included organizations in the public sector. In some cases (see, for example Clarke, 1989; Vaughan, 1996), studies have centered on multiorganizational networks comprised of both public and private organizations. Much of this work is concerned with whether particular organizational structures, cultures, and processes are more prone than others to major accidents and disasters, and if so why. Specific questions addressed in this line of research have included why some organizations are more concerned about safety and better able to translate that concern into effective management of hazards than others and what changes need to occur in organizational practices and organization-environment relations to ensure that organizations operate more safely.

Organizational studies on risk, safety, accidents, and disasters have obvious implications for emergency response research. Much of that research focuses on actual or potential technological failures. Moreover, some studies provide significant insights on organizational behavior during disaster situations. Most important, organizationally-focused research introduces new and potentially useful theoretical perspectives. For example, Mitroff and his colleagues have conducted a considerable amount of research on the origins and management of crises in organizations (see Mitroff, Pauchant, and Shrivastava, 1988; Mitroff et al., 1989; Mitroff and Pearson, 1993). Clarke (1990, 1993) has studied the Exxon oil spill to better understand why organizations fail to plan adequately and respond effectively to major crisis events. Shrivastava (1987) has analyzed the societal and organizational sources of the Bhopal tragedy, including the factors that contributed to the ineffective post-release response.

Much recent work undertaken from an organizational perspective been related to two perspectives on safety: Perrow's (1984) "normal accidents" approach and the analyses of high-reliability organizations that have been undertaken by La Porte and his colleagues (La Porte, 1988; Roberts, 1989; Roberts, Rousseau, and LaPorte, 1993; La Porte and Consolini, 1991). These two analytic frameworks offer differing views not only on what makes organizations safe or unsafe but also on how safe the systems currently used to manage complex technologies really are. In his influential work on normal accidents, Perrow made the case that the manner in which production is structured and managed in some types of organizations that use risky technologies makes those organizations prone to catastrophic failure. Specifically, the potential for disastrous failures is increased when the organizational systems in which those technologies are embedded are tightly-coupled (as opposed to loosely-coupled) and involve complex interactions (as opposed to linear processes). These kinds of organizational systems are unable to correct—or, in many cases, to even detect—small operating problems. System properties further amplify those problems, leading under certain circumstances to a major accident or disaster.

Perrow argues that the inherent weaknesses of these kinds of organizational systems can be overcome to some degree through changing the organization and its relationship with its environment. However, no matter what is done, a small subset of organizational types—and for Perrow this group includes, most notably, nuclear power plants—will remain capable of producing catastrophic losses. Perrow does not claim that such events will be common even in these kinds of systems; many circumstances must conspire to produce a major catastrophe, and most small-scale failures stay small. However, he does contend that political and economic exigencies, such as the need to protect organizational autonomy and prestige, production pressures, and a focus on short-term profits invariably work to lower the priority placed on safety, even in organizations dealing with the riskiest technologies. Unsafe organizational practices can be remedied, but only to a point. In cases in which failures are likely to involve enormous losses, the risks may be judged too great. (For more detail see Perrow, 1984, 1994.)

In contrast, researchers studying what they term high-reliability organizations are considerably more sanguine about the ability of organizations to anticipate and avoid major failures. These researchers have identified practices that they contend enable some organizations that engage in highly risky activities to operate in a nearly error-free fashion despite the kinds of structural constraints Perrow has identified. As examples of these organizational successes, they include the safety records of U.S. aircraft carriers, air traffic control, electrical power grids, and (in sharp contrast to Perrow) nuclear power plants. Organizations achieve high reliability through strategies that include continually searching for potential problems, training personnel extensively, reviewing how future crises should be handled, building redundancy into operations, learning from mistakes, developing organizational cultures that reinforce reliable performance, and decentralizing operational authority when a crisis threatens. According to this perspective, while few organizations actually succeed in performing consistently at very high levels of safety, the fact that some do is evidence that even highly complex, risky technologies can be managed. (For other discussions of how these two traditions differ, the validity of their claims, and empirical studies undertaken from both perspectives see Sagan, 1993; Clarke, 1993; and a paper symposium in the *Journal of Contingencies and Crisis Management* Vol. 2 No. 4, 1994, in which the proponents of the two approaches confronted one another directly.)

Unfortunately, despite their clear relationship to one another, there has not been much connection or cross-fertilization between disaster research and organizational scholarship on disasters and risk (see, for example, Clarke and Short, 1993; Vaughan, 1999; Clarke, 1999). Even though disaster researchers and organizational risk researchers frequently study the same kinds of crisis events, they often appear to be unaware of one another's work. Disaster research tends to be much more descriptive and applied than work in the organizational field, and it has largely failed to employ the theoretical models that organizational scholars have formulated. Similarly, organizational researchers who focus on risk and disaster have failed to take advantage of both the large body of data and the insights developed by researchers in the disaster field.

Mass Media Organizations and Disaster

For many people and in most situations, the mass media are the most salient source of information on hazards and disasters. In large measure, people learn what they know about disasters from the mass media. In light of their pervasiveness and societal importance, it is rather surprising that the activities of mass media organizations in disaster situations have received so little attention from researchers. A 1980 National Academy of Sciences report entitled *Disasters and the Mass Media,* which contained a series of reports from a workshop held in 1979, was the first systematic attempt to focus on media performance during disasters. That report highlighted several important roles the media can play with respect to hazards and disasters, including educating the public about hazards, disseminating disaster warnings, reporting on disasters and their impacts, providing information on available sources of disaster assistance, and coordinating with government agencies and other emergency response organizations. The report also pointed out that prior to that time little systematic research had actually been conducted on how the media perform these functions in disaster situations.

Interest in media operations during disasters has increased significantly, particularly in the last 15 years. In an earlier review, Quarantelli (1989b) could identify only two media studies that had been conducted before the late 1960s: a 1956 master's thesis that analyzed letters to the editor in a newspaper following the 1953 Waco, Texas, tornado and a 1964 Disaster Research Center study on the operations of a radio station during a forest fire near Santa Barbara, California. A literature review conducted by Disaster Research Center in the mid-1980s (Friedman et al., 1986) identified only 26 studies on the mass media's performance during disasters in the English language literature. Since that time, the number of research articles and reports has expanded considerably, driven by large-scale disasters that became major media events, such as Bhopal, Chernobyl, and the Exxon oil spill. There has also been a growing interest in studying disaster reporting on the part of communications researchers. Research on disasters and the media can be categorized into four general areas: studies on media framing of hazards and disaster events; the media as a source of warnings and other types of disaster-related information; media newsgathering practices during disasters; and media emergency preparedness and response.

The bulk of the disaster-related media research conducted to date focuses on the manner in which the media frame disaster reports. Influenced by analyses of the production of news as a social process (Altheide, 1976; Tuchman, 1978; Altheide and Snow, 1979; Gitlin, 1980; Gans, 1980), this research relates the content of media reporting on disasters to a variety of organizational and contextual factors. These include the organization of routine news production, beliefs within media organizations about what constitutes news, and typical media strategies for "packaging" stories (Gamson and Modigliani, 1989). Disaster reporting is also characterized as influenced by news sources and their interests, as well as by broader cultural images of disasters, their causes, and their consequences.

Since the inception of mass media studies on disasters, researchers have pointed out that such events are enormously attractive to news organizations because they fit squarely within the parameters of what makes a "good story"—action, visual impact, and human drama. Unfortunately, however, disaster news coverage is generally less than adequate along a number of dimensions. For example, reporters often lack an understanding of the scientific and policy aspects of the stories they are covering (Friedman, 1989). Although not all studies fit this pattern (see, for example, Goltz, 1985), some researchers also contend that the media tend to reinforce myths about disaster behavior, such as the idea that looting inevitably occurs in disaster situations, perhaps because media personnel themselves subscribe to those myths (Fischer, 1998). Moreover, reporting tends to adopt a "command post" point of view, framing the disaster event from the perspective of government officials and established institutions (Quarantelli, 1981b; Wenger and Quarantelli, 1989).

Research on disasters that became major media events illustrates the problematic aspects of disaster reporting. One such shortcoming is the media's tendency to focus on disasters as isolated events. For example, Wilkins (1987, 1989) analyzed U.S. news reports on the 1984 Bhopal catastrophe, focusing on the ways Bhopal-related stories were handled by wire services, news magazines, the "prestige" press, and television. All four media took an overwhelmingly event-oriented position on the story, providing details on the release and its immediate impacts without delving into either the broader societal causes of the accident or its long-term health and environmental consequences. Few stories suggested a link between Union Carbide's Bhopal plant and its facility in Institute, West Virginia. Prestige publications like the *New York Times* and the *Washington Post,* as well as the news magazines, showed more of a tendency than other media to place the disaster in a socio-political context. However, overall "reports consistently decontextualized the event . . . [and] fragmented the event in a more profound way, for linkages between what was happening in India and what could—and eventually did—happen in Institute did not permeate media coverage of Bhopal" (Wilkins, 1989: 27).

Smith (1992) studied the news-production process in three major disaster events: major forest fires in the Yellowstone National Park area in 1988, the Exxon oil spill, and the Loma Prieta earthquake. His research highlights the ways in which the cultures and work practices of media organizations—for example, the focus on dramatic visual content and the need to fix blame for disastrous events—shape reporting, as well as the ways news stories can perpetuate myths about disasters and their causes.

The relationship between disaster reporting and broader cultural assumptions and values is another common theme in studies on media framing. Patterson (1989), for example, found that network news reporting on the 1986 Chernobyl disaster reinforced Cold War and Reagan-era images of the Soviet Union, depicting the Soviets as inept at plant design, unconcerned about human lives and safety, and secretive and deceptive. Images of the power-plant accident were conflated with images of nuclear war. Once again, the effect was to decontextualize the accident and to construct a myth about the event centering on "the integrative propaganda of the superiority of American technology disconnected from the risks such technology had just as obviously brought to the Soviets" (Patterson, 1989: 133). (See also Nimmo, 1984, on network news construction of fables about technology and vulnerability following the Three Mile Island nuclear accident. For other good overviews of the role of the media in framing disaster- and risk-related information, see Balm, 1993 and Dunwoody, 1992.)

A second major topic in research on the media in disasters involves their role in disseminating hazard-related information to the public. The media are among the public's most important sources of information on long-term hazards, imminent disaster threats, and recommended self-protective actions. As noted above in our discussion of protective responses, their role is particularly prominent in the disaster warning process. Chapters by Mogil (1980) and Carter (1980) in the National Academy of

Sciences report on disasters and the media focus on linkages between the media, the National Weather Service, and other emergency agencies during the pre-impact warning period. A number of subsequent studies have looked at the ways community residents use disaster warnings broadcast by the media and the factors affecting willingness to comply with media-disseminated warnings. Ledingham and Masel Walters (1989) found that the media, particularly television and radio, were important and credible sources of information when hurricanes threatened Galveston, Texas. Similar results were found for communities facing volcanic hazards immediately before and long after the 1980 Mt. St. Helens eruption (Perry, Lindell, and Greene, 1981; Perry and Lindell, 1989).

In the only study of its kind in the literature, Beady and Bolin (1986) analyzed the operations of African-American-oriented news media in Mobile, Alabama, during Hurricane Frederic in 1979, particularly their role in warning the public. Television was relied upon most prior to disaster impact and was considered the most credible information source. Despite the fact that almost all survey respondents had received the hurricane warnings, only 23 percent chose to evacuate. Compared to white-owned and -oriented news organizations, media serving the African-American community were generally found to possess fewer resources, such as emergency generators and alternative transmission towers, that would help them continue operating in a disaster situation. This finding suggests that media targeting minority audiences may be especially vulnerable to disaster-induced damage and disruption.

Research has also focused on the public's use of the media to obtain hazard-related information during non-disaster times. Nigg (1982) and Turner, Nigg, and Heller-Paz (1986) studied the ways in which media reports on the earthquake hazard, including presumed earthquake precursors like the Southern California Uplift, influenced the public's perception of the threat. Their research also focused on which media people use most often for earthquake information (television and newspapers rated highest) and on intergroup differences in media use. In a study discussed above, Mileti and his colleagues (Mileti et al, 1993; Mileti and Darlington, 1995, 1997) assessed how a newspaper insert providing detailed information on the earthquake hazard in the San Francisco Bay Area affected risk perceptions and preparedness behavior among households, governmental organizations, and private businesses. Their data suggest that the insert had a positive impact on residents' knowledge about potential seismic impacts, although there was significant variation in the kind of information they were able to recall. The brochure also appeared to have a positive impact on the adoption of seismic mitigation and preparedness measures.

Other studies have focused more directly on how media organizations gather and disseminate news in disasters. These events pose distinctive challenges for news organizations because they often take place without warning, can occur in relatively inaccessible areas, and can disrupt the communications linkages needed for media operations. Reporters can find themselves covering events that are unfamiliar and therefore difficult to understand and explain to the public. The classic example of this situation was the partial meltdown at Three Mile Island, where one of the questions following a lengthy technical explanation of nuclear plant operations is reputed to have been "What's a valve?" In a case study on media activities following Hurricane David in Dominica that typifies the problems of disaster reporting, Rogers and Sood (1981) found that news organizations were forced to resort to unconventional, improvised communications channels to obtain and exchange information. In the initial stages of the disaster, reporters were unable to identify reliable sources for information, and they were much less able to confirm the information they had obtained than was typically the case during normal operations. Quarantelli (1991) has noted that disasters typically afford reporters more autonomy in their activities than they normally enjoy at other times. He also found that, like other organizations affected by disasters, media organizations undergo considerable structural adaptation when responding during disaster situations, particularly in their internal division of labor. For example, it is common for parts of news organizations that normally perform other tasks to become involved in newsgathering and reporting during disasters.

In the aftermath of a disaster, one of the things the public wants most is information about the severity of the event. The general tendency in early media reporting is to overestimate deaths and injuries. Rogers and his colleagues (1990) explored this pattern in a study of media operations following the 1989 Loma Prieta earthquake, a major media event in which early estimates set the death toll as high as 370 (in all, between 62 and 65 deaths were judged to have been due to the earthquake). They attributed these overestimates to the widespread impact of the event, conflicting severity estimates from affected jurisdictions, the lack of a single, authoritative source for information on the number killed, and the ambiguity regarding how many people were killed by the collapsed Nimitz freeway. Because there was so little authoritative information in the initial period after the earthquake struck, reporters turned to unofficial sources and eyewitnesses for their initial estimates. News media reporting during disasters is affected both by the ongoing pressures that accompany the routine production of news and by the special pressures disasters introduce. Competition is fierce in the media industry. Holton (1985) observed, for example, that tensions typically exist between local and out-of-town media, among the various types of media, between providers of the same media services, and among different programs and correspondents. Although media organizations generally cooperate during disasters, there is also increased competition for breaking news.

Because the number of news organizations that are capable of fielding news teams has increased so dramatically in recent years, media convergence at disaster sites is now commonplace. The Federal Emergency Management Agency's development of the "joint information center" as a way to provide authoritative information to reporters is one effort to cope with the demand for news (Holton, 1985).

Finally, a very limited amount of research has been done on emergency preparedness and response procedures within mass media organizations themselves. Studies that have examined these issues generally have found media levels of preparedness to be poor (Quarantelli, 1991). Rogers and his colleagues (1990), in their study of the Loma Prieta earthquake, noted that television and radio stations in the Bay Area were unprepared for disasters, and that many were hard hit in the earthquake, in part because of their failure to undertake mitigative actions. For example, a number of radio and television stations throughout the region lost broadcasting capacity when offsite electrical power sources failed. Newspapers were also found to be "generally unprepared to operate under the emergency conditions of the disaster" (Rogers et al., 1990: 36). For example, like their counterparts in the electronic media, two major papers in the Bay Area, the San Francisco *Chronicle* and *Examiner,* experienced severe operational problems due to the loss of power.

Wenger (1985) has pointed to two competing images in the literature (we might call them social constructions) on how the media operate during disasters: the media as foe and the media as friend. On the one hand, disasters are framed by news organizations in ways that can be misleading. By focusing on disasters as events divorced from their contexts, the media can reinforce the public's oversimplified views of why disasters happen. The media also can convey to the public erroneous impressions about the magnitude and even the location of disaster damage. For example, San Francisco was characterized as virtually in ruins following the Loma Prieta earthquake, when in fact the city was only selectively damaged, and the site of the most severe destruction was in Santa Cruz, 70 miles away. To the extent that the news media perpetuate myths about disaster behavior, they can convey unrealistic impressions about disaster-related needs and problems. In turn, this leads both the public and crisis decision makers to worry about the wrong things. Finally, overreliance on a limited group of official news sources that typically have strong vested interests can cause news organizations to slant their reporting with one-sided views on hazard-related issues.

On the other hand, that the media also make a strong positive contribution in disaster situations is inarguable. Effective warnings broadcast through the media are widely credited with playing a major role in the reduction of casualties from hurricanes, tornadoes, and floods. When they are able to convey timely and accurate information on threats, the media can save lives. More than half-a-million people were safely evacuated from southern Florida before Hurricane Andrew struck in 1992. Without extensive media involvement in disseminating warning information this would have been impossible, and the death toll from the storm would undoubtedly have been much higher.

By reporting extensively on disasters and the damage they create, the media can help speed up assistance to disaster-stricken areas, and post-disaster reporting can also provide reassurance to people who are concerned about the well-being of their loved ones. Good science reporting can educate the public about hazards, and in-depth stories can help provide the basis for informed hazard-reduction decisions. Scientists who understand how the media operate and who develop good working relationships with media organizations can become highly credible information sources to whom the public turns when trying to decide what to do in disaster situations. Prominent examples of such scientists include Neil Frank, a former chief hurricane forecaster at the National Weather Service who went on to become a media figure, and earthquake experts Lucy Jones and Kate Hutton at the U.S. Geological Survey/California Institute of Technology offices in Pasadena. Thus, as Scanlon and his colleagues (1985) have emphasized, while the media can often create problems for public officials—for example, by converging en masse to disaster-stricken areas—they can also be a major asset, particularly in light of their crucial role in communicating with the public, and efforts should be made to include media organizations in emergency planning activities.

In his review of research on mass communications in disasters, Quarantelli (1989b) noted the need for research on a number of media-related topics. Among these are studies on the gatekeeping function of the media during disasters; structural alterations that media organizations undergo in disaster situations and the ways in which factors such as organizational size affect those changes; and differences in local-level and national-level reporting on disasters. He also called for research to better understand how mass media networks and informal social networks such as those comprised of friends, neighbors, and relatives interact with and supplement one another in disasters. Finally, Quarantelli pointed to the need to study the impacts of the large-scale and rapid changes that are occurring in media and communications technology.

INTERORGANIZATIONAL AND COMMUNITY RESPONSE

In the 25 years since the first assessment of research on natural hazards, research on multi-organizational response activities has yielded important conceptual, methodological, and analytic contributions despite the limited number of studies that have been conducted. Among the most important of these contributions are Drabek's work on emergent multi-organizational networks (EMONs) (Drabek et al., 1982; Drabek, 1983b, 1985) and Kreps's research program on organizing for disaster response (Kreps, 1985, 1989; Bosworth and Kreps, 1986; Kreps and Bosworth, 1993).

The study of emergent multi-organizational networks in the post-disaster period is not an entirely new development in the field. Some work had been done on the topic prior to the time Drabek and his colleagues began their research, although much of that earlier work concentrated more on the recovery period than on response activities (e.g., Taylor, 1976; Ross, 1980). By contrast, Drabek's research dealt more directly with the emergency period, focusing specifically on emergent search and rescue (SAR) networks in six major natural disasters. The EMONS they studied were comprised of a mix of public and private organizations and volunteer groups. The networks were conceptualized as emergent, not because they were comprised of emergent groups (although such groups were involved) but because the network relations that developed were emergent rather than planned prior to disaster impact. That is, contact, communication, and other dimensions of interorganizational relations developed and evolved during the response period.

The prevailing patterns in SAR response in the six disasters studied included localism, lack of standardization, unit diversity, and fragmentation. Consistent with other studies in the literature, Drabek found that

SAR was carried out by local community organizations. The networks that developed in the six communities differed in terms of which organizations were central and how those organizations were structured. Moreover, the mix of organizations that were involved was not consistent across events; and the networks lacked cohesiveness on several dimensions, particularly interorganizational communications (Drabek et al., 1982; Drabek, 1983b, 1985).

Among the most significant problems noted for these networks were difficulties with communications, ambiguity of authority, and poor utilization of special resources. For example, communication with the public must pass through the news media, and as noted above, the public's response can be problematic if the media incorrectly transmit information or promote an interpretation of events that is based upon disaster myths. However, EMONs were poorly prepared to deal with the media.

The Drabek study suggested that coordination problems inherent in EMON operations can be overcome when there is high domain consensus among organizations, which occurs when each organization understands the purpose of the network, its own role, and the roles of the other organizations. Emergency response is also more effective when there is an identified leader with both positional power (especially legitimate authority) and personal power (e.g., expertise) operating through a central coordinating mechanism such as an emergency operations center. Finally, responses are more effective when the responding units have frequent interaction with one another prior to a disaster, especially when they periodically participate in joint exercises. This is because it is inherently very difficult to develop domain consensus and authority structures and to acquire specialized resources during large-scale, rapid-onset events. One limitation of this study is that it focused on only one emergency-period task—search and rescue—so the generalizability of its findings to other activities remains to be assessed.

Kreps's research on organizations and emergence during the disaster response period used data from the archives of the Disaster Research Center. Those data consisted of descriptions of organized action that were contained in just over one thousand organizational interviews conducted between 1964 and 1972 on 15 disaster events. The more than 400 instances of organized action described in those interviews were classified in terms of their underlying structural properties in order to better understand the range of organizational forms that characterize the emergency response period.

Kreps's work represents the most sustained and detailed effort to date to link the study of disaster response activities to general social theory. The assumption on which this theoretical project is based is "that forms of association enacted during disasters reflect sequences of 1–4 basic elements of social organization: domains (D), tasks (T), human and material resources (R), and activities (A)" (Kreps, 1984: 315). A logical combination of these four elements of organization—D, T, R, and A—yields 64 possible organizational forms. It is important to note that the units of analysis for Kreps's taxonomy are organized responses, rather than organizations. The application of this framework to the behavior of populations of social units that were active during the response period in the disaster events studied revealed a wide array of forms of organizing, ranging from structured, means-ends-oriented action on the one hand to almost pure collective behavior on the other.

Kreps's research also focused on patterns of role enactment in the disaster response phase. Performance of disaster roles was shown to vary, ranging from conventional and routine—as in the case when familiar, formally-designated disaster responsibilities are carried out—to improvised. Stable role enactment was more likely to take place when participants got involved rapidly in less severe disasters; when actors had some prior experience carrying out their roles; when the roles in question required significant knowledge; when the organizations involved were expected to have disaster-related responsibilities; and in more formally-organized, more self-contained, less complicated, larger responses in metropolitan communities. To the extent these conditions were not present, role change and improvisation were more likely to occur. (For more detailed discussions see Kreps, 1985, 1989; Bosworth and Kreps, 1986; Kreps and Bosworth, 1993.)

In a study focusing primarily on organizational effectiveness in the dissemination of warning messages, Sorensen, Mileti, and Copenhaver (1985) reviewed the literature in an effort to identify the correlates of cohesive emergency responses, both for individual organizations and interorganizational networks. Based on that review, they concluded that interorganizational cohesiveness during the response period is influenced by several factors. First, there must be domain consensus, or a clear understanding of the responsibilities of each organization in the network, as well as mechanisms for resolving disputes among organizations. Additionally, organizational legitimacy, resource adequacy, and organizational willingness to give up some autonomy for the good of the overall response system also encourage a more cohesive response. Finally, cohesiveness is more likely when there are high-level communications linkages among organizations, clearly established authority structures, clarity with respect to lines of interorganizational contact, and shared knowledge of the way the system is supposed to operate. These factors predicted response effectiveness in a subsequent study of response planning activities at a nuclear power plant.

In an interorganizational study with a different emphasis, Clarke (1989) focused on the immediate and longer-term response issues organizations faced in dealing with an episode of PCB contamination at a state-owned office building in New York. Part of the problem in responding to that event was that responding organizations were unclear about the level of hazard the contamination presented and what should be done about it. Moreover, no organization or group of organizations had a clear mandate to be involved in emergency response and clean-up activities. Clarke's analysis of the situation used garbage-can theory (March and Olsen, 1979) to explain the behavior of actors in the interorganizational network that eventually emerged to deal with the situation. Rather than making decisions rationally, as some theories of organizational behavior would contend, the organizations faced with handling the PCB incident dealt with the problem through negotiation and occasionally—if they were able—through simply exiting the situation. At least initially, connections among response organizations were loosely-coupled, lines of authority and responsibility were ambiguous, and there was considerable conflict. Clarke argues that the garbage-can approach is "especially useful for drawing attention to organizational behavior under ambiguous conditions, i.e., situations in which goals are unclear, technologies are ill defined, and rights to participate in major decisions are in flux" (1989: 174).

While natural disaster situations usually do not fall into this category, crises involving technological hazards often do; Love Canal and Times Beach come immediately to mind. Clarke's research also calls attention to the fact that a decision about whether or not to consider an event a disaster can in some instances be a highly contested issue. When an entire state is flooded, as happened in Iowa in 1993, no one would disagree that a disaster has occurred. However, in other situations, the "disastrousness" of an event may itself be in dispute, and this lack of consensus will almost certainly have ramifications for interorganizational relationships.

Research conducted during the late 1970s and early 1980s by the Disaster Research Center focused on the interorganizational and community response to 20 major emergencies involving hazardous chemicals, including chemical plant explosions and airborne chemical releases. The study found that in general the response to such events was marked by poor coordination among responding organizations and between the public and private sectors. Initial responders arriving on the scene often lacked information about the chemical hazards involved and were confused about the degree of danger these emergencies posed and what actions to take. Problems with obtaining and disseminating accurate information were common, and the involvement of extra-community organizations and high levels of convergence further complicated response efforts.

Reports from this study stressed the importance of situational factors in either facilitating or hindering response efforts. Factors such as the time of day, day of the week, whether an emergency occurred during daylight or nighttime hours, and the jurisdiction in which it occurred were shown to have a significant impact on response effectiveness. For example, the well-handled response following the Mississauga, Canada train derailment in 1979, one of the events included in the study, was seen as resulting not only from pre-planning but also from favorable situational contingencies (Gray, 1981).

Studies that compare how different local jurisdictions have responded to the same disaster event are rare in the literature. In one such study, Stallings and Schepart (1987) studied governmental response patterns in two different communities that were struck by the same tornado. In one community, a locally-centered, locally-directed response developed along the lines that are typically observed in U.S. disasters. In the other, state government became directly involved in handling the response—even to the extent that the governor personally oversaw many of the activities that were carried out in the community. Differences in the way the response was handled were not attributable to disaster event characteristics like severity, since disaster impacts were comparable across the two communities. Rather, response activities were shaped by local governmental structure and the nature of the intergovernmental relationships that had existed prior to the tornado. State involvement in the second community was an extension of cooperative activity that had already taken place in areas, such as city planning, as well as in earlier emergencies.

The 1994 Northridge earthquake struck in a densely-populated part of Southern California and affected a number of separate jurisdictions in three counties. Research by Nigg (1997) and Bolin (1998) compared the manner in which different local jurisdictions responded to the earthquake. Focusing on three heavily-impacted jurisdictions, Nigg (1997) found that those governmental units differed in the extensiveness of their pre-earthquake planning efforts, the ways in which they implemented their plans at the time of the earthquake, and particularly in their relationships with the county, state, and federal levels of government. Intergovernmental relationships during the emergency response period were influenced not only by pre-event planning and prior disaster experience but equally importantly by ongoing patterns of intergovernmental contact, informal relationships among personnel and agencies that were activated to meet emergency needs, and the amount of political power different jurisdictions were able to exercise.

Bolin's (1998) study, which focused on five California communities—Los Angeles and four communities in Ventura County and the Santa Clarita Valley—was concerned mainly with how those communities performed tasks associated with emergency relief and early recovery, particularly the provision of housing for victims. Like Nigg, Bolin found major differences not only in the kinds of problems each community faced in the aftermath of the earthquake but also in levels of community preparedness, overall emergency management capabilities, and the manner in which communities handled relationships with other governmental levels. These differences were rooted in turn in communities' resources and their positions in the intergovernmental disaster management system. For example, one very small unincorporated community had no disaster plan, and the initial emergency response was handled primarily by an informal volunteer network. The community had no pre-established procedures for conducting damage assessments or for requesting assistance through formally-designated channels. Consequently, even though the community had been extensively damaged, outside aid was slow in arriving. At the other end of the continuum, larger and more affluent communities had professionalized emergency management organizations and well-equipped emergency operations centers. Those communities understood intergovernmental disaster response procedures and were familiar with how to access outside response and recovery assistance.

A key insight these studies provide is that, even in a region of the country that is widely recognized as being well-prepared for disasters, individual communities differed markedly in their ability to cope during the emergency response period. The Northridge studies also suggest that, while on average emergency management capabilities may well be improving nationwide, significant intercommunity differences still exist, and researchers need to exercise caution when generalizing about community-level disaster preparedness and response.

Although progress has certainly been made since the time of the last assessment, researchers whose interests center on organizational and community emergency response have also missed some important opportunities. Being able to make cross-community comparisons while holding constant disaster characteristics such as agent type and severity would make it more feasible for researchers to begin to generalize about factors that affect emergency response activities at the organizational, interorganizational, and community levels. However, studies like those carried out following the Northridge earthquake, which involve comparisons and contrasts across multiple local jurisdictions affected by the same disaster event, are all too rare in the literature. Additionally, as suggested by the Clarke PCB contamination study, which employed the garbage-can approach to better understand how organizations seek to define and address (or in some cases to avoid) emergency-related problems, the same general theoretical approaches that are used to study the behavior of organizations during non-disaster times have obvious applicability to the study of disaster response. Yet many disaster researchers have been slow to draw upon insights from the broader organizational literature, and much recent research lacks a sound theoretical foundation. This is perplexing indeed for a field that is fundamentally concerned with organizational issues.

Community-Level Response Issues

Early studies on community response to disasters (Fritz, 1961a; Barton, 1969; Dynes, 1970; Dynes and Quarantelli, 1971) documented a number of changes that occur at the community level in disaster situations. These include enhanced community solidarity and morale, suspension of pre-disaster conflicts, a leveling of status differences, increased levels of community involvement and participation, and shifts in community priorities to emphasize central tasks such as the protection of human life. More generally, the disaster-stricken community has been described as altruistic, therapeutic, consensus-oriented, and adaptive. In a classic statement, Dynes (1970: 84) observed that:

> Disasters create unity rather than disorganization. The consequence of a disaster event on a locality is in the direction of the 'creation' of community, not its disorganization, because during the emergency period a con-

sensus of opinion on the priority of values within a community emerges;
a set of norms which encourages and reinforces community members to
act in an altruistic fashion develops; also, a disaster minimizes conflict
which may have divided the community prior to the disaster event.

As our earlier discussions on individual responses, disaster volunteers, and the provision of emergency aid indicate, research has shown
that disasters result in heightened levels of pro-social behavior. Cooperation and consensus are high during the emergency response phase, self-
interested activity is discouraged, and existing community conflicts are
temporarily set aside. Correspondingly, disaster-specific anti-social behavior, such as the looting that is often anticipated by authorities, the
media, and the public fails to materialize, and the incidence of "ordinary" crimes such as theft and murder declines.

Little of the research that has been conducted on post-impact emergency period activities over the past 20 years contradicts this general
picture. Communities do rise to the occasion when disasters strike, and
the modal pattern reported by researchers is one of heightened community solidarity. However, some recent research indicates that exceptions
to this pattern do exist, and a few scholars argue that the overwhelmingly positive picture that has been painted of the community in the response period may have resulted in a failure to attend to other, more
conflictual processes that also appear to be quite common during the
post-disaster recovery period (see, for example, Stallings, 1988).

As noted above in our discussion of the emergence of groups that
avoided using formal assistance structures and devised alternative sheltering arrangements in Santa Cruz County following the Loma Prieta
earthquake, disasters can become occasions for organized resistance
against established institutional structures and bureaucratic procedures.
In the Loma Prieta situation, pre-existing community conflicts re-emerged
quite rapidly following disaster impact. Ethnic solidarity, which had
served as the basis for mobilization on other issues, such as a strike at a
local cannery prior to the disaster, was a major factor in that resistance
(Simile, 1995).

Other research has focused on the ways in which certain types of
hazard agents have engendered conflict rather than community solidarity. In *The Real Disaster is Above Ground,* Kroll-Smith and Couch characterize the community response to an ongoing underground mine fire in
Centralia, Pennsylvania, as "at wide variance from the typical communal
response to an immediate-impact disaster" (1990: 43). In making the
case that what they term "chronic technological hazards" differ from

natural hazards in their effects, they point to the protracted nature, ambiguity, and uncertainty of these events. As discussed in more detail in Chapter Five, Kroll-Smith, Couch, and others argue that, unlike natural disasters, technological threats do not increase consensus; instead, they erode it. Similarly, Clarke's (1989) study on contamination at a state office building, discussed above, found interactions among response organizations to be conflictual, rather than consensus-driven. The response to the 1989 Exxon oil spill is also depicted in the literature as involving considerable interorganizational and community conflict, rather than the high morale and cooperative spirit that attends many natural disasters.

The point we want to emphasize here is that researchers are increasingly calling attention to the fact that there are situations involving hazard agents in which interorganizational and community consensus are low, rather than high. In such situations, researchers have observed various response-related problems, including lack of consensus on authority and responsibility, uncertainty about which organizations should be involved in the response, ambiguity about what should be done, and even questions about whether it is necessary to do anything at all.

Stallings (1988) has argued that, even if the finding that disasters are characterized by consensus and cooperation still holds, that point needs to be qualified in several ways. First, heightened community consensus is generally characteristic only of the emergency response phase during and immediately following impact; conflict is common both before disaster strikes and during the post-disaster recovery period. Second, generalizations about community consensus and the emergence of pro-social norms are based largely on U.S. society, and we actually know little about how applicable they are to other countries, because so little disaster research has been done in other societal settings. Third, high levels of consensus are probably more characteristic in situations defined as acts of nature than in other kinds of emergencies.

In Chapter One we pointed out that functionalist theory has been the dominant perspective in the field of disaster studies, and we discussed alternative theoretical approaches that are beginning to have an impact on research, including conflict-based perspectives. A conflict-oriented view of disasters interprets the patterns that emerge in the post-impact response period through a different lens. For example, Stallings (1988), following earlier work by researchers such as Hewitt (1983), argues that the suppression of ongoing social conflicts, intergroup struggles, and patterns of dominance during the emergency period can be attributed to two factors. The first is the expansion of direct involvement by the state in allocating resources and managing the response, which is aimed at restoring the market economy as rapidly as possible. The second factor is the temporary compatibility that exists between the interests of dominant and subordinate groups when a disaster occurs. Consensus is, in other words, temporary, provisional, and generated by the same underlying processes that shape social relations during non-disaster times.

It is difficult to say whether community conflict was ignored in previous research, whether it is becoming more evident because different kinds of events are being studied, or whether it has actually increased in recent years. Conflict seems to have become more prominent as a theme as researchers have begun to extend the scope of disaster research—originally studies of sudden-onset, physically destructive events causing immediate casualties and social disruption, such as tornadoes, earthquakes, and explosions—to include slow-onset, physically contaminating hazards that might take decades to manifest their effects, such as exposure to chemical toxins. In marked contrast with the view of natural disasters conveyed in the literature, situations involving these kinds of hazards can often be divisive and conflict-ridden.

However, instances of community conflict in all types of disaster situations may also be increasing, for a variety of reasons. As knowledge about the social sources of hazard vulnerability is becoming increasingly widespread, members of the public are becoming more aware than before of the ways in which the actions of others can cause disasters or make their impacts more severe. An instance of victimization that may once have been seen as resulting from an act of God, the uncontrollable forces of nature, or sheer bad luck may now be seen as having been caused by some party's negligence. These new interpretations can in turn lead to conflict, criticism of organizational performance, and in some cases litigation. Moreover, members of the public may now expect more from government when disasters strike than they once did. Consequently, problems such as traffic jams during pre-impact evacuation and delays in infrastructure restoration and service delivery following disasters may now be judged more harshly by those who are affected. As the widespread criticisms and the subsequent investigation of federal emergency management policies following Hurricane Andrew showed, the public expects government to respond swiftly and effectively in emergencies and has little tolerance when those expectations are not met.

Thus, while the older solidaristic view of community disaster response retains its validity, newer research increasingly recognizes that disasters can engender conflict as well as consensus and that solidarity and conflict often coexist in disasters, just as they do during non-disaster times. Indeed, it has long been recognized that the conception of disasters as "consensus" crises that differ from "dissensus" emergencies, such as riots, is an ideal type, not an exact reflection of reality. As recent research indicates, the emergency period can be marked by competition among organizations as well as by cooperation, and by the formation of internally solidaristic groups that oppose one another or criticize governmental response measures. While sharing a broad consensus on goals, groups can still disagree on how those goals should be achieved. And while emergency situations may well be characterized by heightened morale and moves toward greater inclusiveness, those generalizations are likely to be more true for those who are already well-integrated socially than for marginalized groups.

SUPRA-LOCAL AND NATIONAL RESPONSE

As we saw in the previous chapter, as the focus of research moves from the micro- to the macro-level empirical studies become much more scarce. We noted above that, while states are very important actors in the intergovernmental disaster management system, we currently know very little about state-level preparedness. The same thing can be said for our knowledge concerning the role of state governments in emergency response activities. The lack of focus on state activities is puzzling, since the actions taken by states can mean the difference between a well-run and an ineffective disaster response.

In a study on intergovernmental response operations that emphasized the role of both state and federal response agencies, Schneider (1995) tried to explain why some disasters are handled rapidly and effectively while other disasters elicit a confused and inappropriate response that is evaluated poorly by the public. Her research looked at several natural disasters, including Hurricane Hugo in St. Croix, South Carolina, and North Carolina; the Loma Prieta earthquake in California; and Hurricane Andrew in South Florida and Louisiana. In some of those cases—notably Hurricanes Hugo in St. Croix and Andrew in South Florida—the intergovernmental response was judged to be almost wholly inadequate. In other cases, such as the Loma Prieta earthquake, government performance succeeded in some areas and failed in others. In still others, such as Hurricanes Hugo in North Carolina and Andrew in Louisiana, response and relief operated smoothly and were assessed as successful.

Schneider attributed these variations in response effectiveness to several factors. First, an effective response is characterized by few discrepancies between government's performance, which is based largely on bureaucratic norms, and the norms and definitions of the situation that emerge in the public. In contrast, poorly-managed events reveal large gaps between the actions government takes and the public's collective definitions. Government, in other words, is not perceived as attending to the needs people consider most important, or is not acting swiftly enough. At the rare extreme, represented by the situation on St. Croix following Hurricane Hugo, government is incapable of acting, and the norms that emerge in the disaster-stricken area permit anti-social and destructive behavior.

A second and related point is that response activities tend to be judged more positively when they develop from and are largely under the control of lower levels of government. Conversely, poor evaluations of government performance are more likely in situations where there is confusion about which governmental levels are responsible for which tasks. The public and the media become aware of this when authorities procrastinate, equivocate, or reverse their decisions, or when there is a lack of coordination among agencies and levels of government. The most visible indication of poor performance by local jurisdictions occurs when a higher level of government intervenes to assume major responsibility for emergency operations.

Third, Schneider argued that both actual and perceived government effectiveness are related to three general factors: disaster magnitude; the extent to which governmental agencies prepare effectively for disasters; and the public's capacity to cope with disaster impacts. The worst-case situation—the disaster that involves the most serious gaps between bureaucratic performance and public expectations—is one in which disaster impacts are very severe, governmental responders are unprepared, and the public lacks experience with the type of disaster that has occurred and is unable to mobilize its own resources effectively. This is also the kind of situation in which a high degree of federal intervention in the performance of response- and early recovery-related tasks is most likely.

Schneider presented an interesting analysis of the intergovernmental emergency response system, how it works during disasters, and what can happen when the system encounters unexpectedly large demands or fails to assess needs in the same way other participants do. She also advanced a testable set of hypotheses on what accounts for governmental effectiveness during the emergency response period. However, while not completely ignoring the political factors that influence both response activities and public perceptions of response effectiveness, as well as the role of the mass media in shaping public perceptions—a key topic discussed above—her formulation tended to downplay their significance. It is important to consider the roles of politics and the media because governmental confusion and intergovernmental conflicts in the aftermath of disasters may be caused as much by pre-disaster political factors as by the kinds of factors she identifies. As she acknowledged but did not really emphasize, concern about the federal response to Hurricane Andrew was related in part to the fact that it occurred three months before the presidential election, and Florida was an important state in that election. Similarly, the intensity and scope of federal involvement in California following the 1994 Northridge earthquake can only be understood in the context of presidential electoral politics and of the federal government's desire to use the disaster to revive the state's stagnant economy. When supra-local levels of government intervene aggressively following a major disaster, it is not always because actions initiated at lower governmental levels have been ineffective. Rather, such intervention may originate in the need to avoid criticism for not being sufficiently proactive, to claim credit for response activities that are proceeding well, or (though those involved would never admit it) simply to grab headlines.

Comparative studies like those discussed above are all too rare in the literature. One notable and virtually unique exception is Comfort's detailed comparative research on response activities following eleven earthquakes that occurred in nine different countries, including the U. S., between 1985 and 1995 (Comfort, 1999). That study, which characterizes multi-organizational response networks as emergent, self-organizing systems, traces why those systems were adaptive and effective in some cases but not in others. More systematic research of this kind, both within and across societies, is clearly needed.

ISSUES FOR FUTURE RESEARCH

Throughout this chapter we have pointed to various areas in which our knowledge is still seriously lacking. There is a dearth of information on interorganizational and intergovernmental relations during the emergency response period and on the factors that make such coordination successful. EOCs are an important mechanism for achieving interorganizational coordination because they are the hubs out of which emergency operations are coordinated. However, very little research exists on how local EOCs are organized, what their capabilities are, or how they function in emergency situations. This clearly is another area in which cross-community research is needed.

The kind of large-scale, systematic, comparative research we advocate is currently conducted infrequently, largely because project budgets for disaster studies tend to be modest. However, without such research, study results will lack generalizability, and important questions will remain unanswered. An obvious method for achieving this goal is to increase the number of projects in which a single research team uses a single instrument to study multiple disasters taking place in different communities. One disadvantage of this approach, however, is that in the absence of significant enhancements in budgets for conducting disaster research it could tend to concentrate funding in a smaller number of institutions. An alternative approach to achieving desired inter-community comparisons would be for different investigators to coordinate their research efforts through the use of common measures and scales. This approach, which we have also advocated for research on households and other social units, would help ensure that research findings are comparable and cumulative.

There also may be a tendency in the field and on the part of some funding agencies to believe that, because certain topics (e.g., myths about disaster behavior, or emergent groups and networks, or the response activities of emergency-relevant organizations) have been studied in the past, we know enough about the processes involved that it is not necessary to conduct further research. Such assumptions are unwarranted, since as social scientists we also know that social change is continuous and that human and organizational behavior are context-dependent. Many older research findings may be quite robust, but we will not know that unless we continue to subject those findings to empirical examination. And there are many important areas—including the entire field of emergency management policy and practice—in which change has been so profound that older research findings must be reconfirmed.

The questions we ask and the studies we undertake also must take into account new theoretical developments, or else the field of hazards research will stagnate. As new perspectives are developed in fields like communications studies, psychology, and organizational sociology, we must be ready to adapt them theoretically and apply them empirically. And we must continue to revisit questions that are central to our understanding of how individuals, households, and organizations respond in emergencies. Similarly, if we fail to continue our inquiries into basic processes of communication, coordination, and decision making in crises, we run the risk of making policy recommendations that miss the mark because they are based on outdated knowledge.

Factors Influencing Disaster Preparedness and Response

THE PRECEDING CHAPTERS HAVE been organized around preparedness and response as temporal disaster phases and around the social units and levels of analysis on which research literature has been conducted. In Chapters Five and Six we take a different approach. Chapter Five synthesizes research findings and identifying factors that act as important influences on behavior across social units and across the two disaster phases. Chapter Six incorporates broader contextual factors into our discussion by identifying—or, where evidence is lacking, by speculating on—the role that more macro-social forces such as governmental structure and culture play in shaping the ways in which individuals, groups, formal organizations, and societies prepare for and respond to disasters.

FACTORS ASSOCIATED WITH THE EXTENT AND QUALITY OF PREPAREDNESS AND RESPONSE ACTIVITIES

In this section we discuss research findings suggesting how preparedness and response activities vary as a function of several different sets of factors: perceptions

157

of hazards and hazard adjustments; disaster experience; socio-cultural factors, including ethnicity, minority status, and language differences; social networks and interorganizational linkages; economic resources, such as income and wealth; gender; and related socio-demographic influences. Although we generally discuss these topics separately, they are of course interrelated. For example, ethnic groups differ from the majority population and from one another in the nature and extent of their social networks, their access to hazard information, and their political power and economic resources. Indeed, as we emphasize throughout this book, a major problem with the research that has been conducted to date is that too little of it exists to make it possible to disentangle the independent and interdependent effects that different factors may have on patterns of preparedness and response. Despite all the empirical research that has been conducted since the inception of the field of hazards research, we really are only in the preliminary phase of discovering and understanding these complex relationships.

Risk Perceptions

It has been known since before the first assessment that people may be aware of a hazard but fail to personalize the risk (Mileti, Drabek, and Haas, 1975). This principle has been replicated repeatedly in the past 25 years. In the 1970s, Jackson and Mukerjee (1974) found that 86 percent of their respondents had experienced an earthquake, and nearly half (43 percent) thought that another earthquake would occur in the next few years, but only about one-third expected to be affected personally. Further, among those expecting damage from a future earthquake, nearly half thought that damage would be slight or had no clear idea of how much damage they would incur. More recently, Mileti and Fitzpatrick (1993) found that about 80 percent of their respondents believed they would experience a Parkfield earthquake, but only about one-third thought it would harm them, their families, or their property.

Although the constructs and measures that have been used in different studies have varied considerably, personalizing risk appears to be an important link between knowing about a hazard and taking self-protective action. Most but not all studies have found significant correlations between risk perceptions and the willingness to adopt hazard adjustments (Lindell and Perry, 2000). For example, Turner, Nigg, and Heller-Paz (1986) found seismic preparedness was significantly higher among those who had heard, understood, and personalized the risk than among those who had not. Again focusing on the earthquake threat, Showalter (1993) found statistically significant effects of concern about threats of death and injury on all protective responses except insurance purchase. And findings from a number of other studies (see, for example, De Man and Simpson-Housley, 1987; Mileti and O'Brien, 1992) have also been consistent with this general pattern.

Once again, however, the literature contains contradictory findings. For example, Russell, Goltz, and Bourque (1995) found that a high level of personal concern—measured as having frequent thoughts about earthquakes—significantly predicted mitigation and preparedness behaviors in only a small proportion of their analyses. Jackson (1977, 1981) found that expectations about future earthquake losses did not predict the adoption of preparedness measures, and Mileti and Darlington (1997) and Lindell and Prater (2000) also found no evidence of a relationship between personal concern and preparedness.

Other research suggests that risk perception, which is often defined in terms of individuals' expectations about both the probability and severity of disaster impacts, may be less important than hazard intrusiveness—defined as the frequency of thinking about, discussing, and receiving information about a hazard—in predicting the adoption of mitigation and preparedness measures. In other words, while people may be generally concerned about a hazard, particularly after a disaster event or after receiving hazard-related information, the salience of the hazard in people's lives may well decline in the face of other more daily concerns unless a potential threat is re-emphasized continually through interaction. The importance of hazard intrusiveness as a factor explaining preparedness behavior can be seen in Lindell and Prater's (2000) finding that intrusiveness significantly predicted hazard adjustments even when risk perception did not, and also that it predicted adjustment better than hazard experience or a range of demographic variables.

Perceptions of Hazard Adjustments

Variations in preparedness are also traceable to variations in the public's propensity to adopt different preparedness measures. Clearly, these differences are related in some degree to variation in public awareness. For example, research conducted in the 1970s revealed that a significant proportion of those who had not purchased hazard insurance were unaware of its availability (Sullivan, Mustart, and Galehouse, 1977; Kunreuther et al., 1978). Although awareness of recommended vulnerability-reduction measures for various hazards has undoubtedly grown since the time of those studies, lack of knowledge clearly remains a factor in the adoption of hazard adjustments, particularly for those who may not have received or understood hazard-related information. In addition to differences in levels of awareness, rates of adoption for different preparedness measures are also likely to be influenced by the time, expense, and effort involved in adoption, as well as to the extent to which measures are seen as serving multiple uses (Edwards, 1993; Russell, Goltz, and Bourque, 1995). Research by Davis (1989) suggested the importance of four attributes: awareness, perceived effectiveness, cost/effort, and the knowledge required to implement the measure. Other influential factors may include whether a particular protective measure is mandatory or voluntary, whether it can be done by the average layperson or needs to be carried out by a professional, and whether it can be performed on a one-time basis or needs to be carried out repeatedly. (For other discussions on variations in rates of adoption for different mitigation and preparedness measures see Davis, 1989; Farley et al., 1993; Mileti and Fitzpatrick, 1993; and Mileti and Darlington, 1997.)

Other data on this topic come from Lindell and Whitney's (2000) study in which respondents from Southern California were asked to report whether they had adopted each of 12 seismic hazard adjustments—two mitigation measures, nine preparedness actions, and purchasing insurance. They were also asked to rate each adjustment in terms of three hazard-related attributes (efficacy in protecting persons, efficacy in protecting property, and usefulness for purposes other than earthquake protection) and five resource-related attributes (cost and requirements in terms of special knowledge and skill, time, effort, and the need for cooperation with others). Study results showed substantial differences in the manner in which respondents rated the different adjustments. All three hazard-related attributes were significantly correlated with both intention to adopt the 12 measures and with actual adoption, but none of the five resource-related attributes were related.

Despite the fact that a considerable amount of research has focused on the adoption of mitigation and preparedness measures, we still know relatively little about how people perceive different measures and what makes some adjustments more attractive to them than others. This is surprising in light of theoretical approaches such as Fishbein and Ajzen's theory of reasoned action (1975), which suggests that people's behavioral responses—in this case, their adoption of loss-reduction measures—may well be more highly correlated with their belief about the behaviors (that is, about the preparedness activities themselves) than with their beliefs about the situation that motivated the behaviors in the first place. In other words, rather than being shaped by more commonly-investigated factors such as risk perception, experience, or socio-demographic variables, the propensity to carry out different self-protective actions may also be influenced in important ways by levels of awareness and attitudes toward those actions, such as their situational appropriateness, probable effectiveness, cost, and ease of implementation.

Disaster Experience

After conducting studies for well over a generation in various community and hazard settings, researchers have been able to isolate relatively few factors that seem to be major predictors of how preparedness and response activities are undertaken across various social units. One which has been identified is prior disaster experience. Even here, however, the literature contains conflicting findings, and there is a great deal we still do not know.

In general, the research literature suggests that prior experience engenders higher levels of preparedness and more effective performance during the response period, largely because it leads to greater awareness of the consequences of disasters and the demands that disasters generate. Evidently, adaptation and learning take place as a result of involvement in disaster situations, so that threats are taken more seriously and necessary tasks and activities are carried out more effectively in subsequent crises.

At the individual and household levels, many studies have shown experience with actual events has a generally positive impact on the willingness to prepare for future disasters (Lindell and Perry, 2000). Focusing specifically on the earthquake hazard, for example, seismic emergency preparedness has been found to be directly related to the number of earthquakes experienced (Russell, Goltz, and Bourque, 1995) and to experiencing earthquake losses either to oneself or to close associates (Jackson, 1977; 1981; Turner, Nigg, and Heller-Paz, 1986). Dooley et al. (1992) found evidence suggesting that experiencing an earthquake that was considered frightening indirectly affected seismic preparedness.

A number of studies suggest that, when individuals have been through one or more similar disasters, they have a greater tendency to believe disaster warnings and consider themselves personally at risk when they receive warning information. They are also more likely to have developed some sort of a plan for responding to disaster situations, which in turn raises the probability that they will act when the need arises (Lindell and Perry, 1992). More specifically, Riad, Norris, and Ruback (1999) found that having been involved in previous evacuations was the single strongest factor predicting household evacuation responses in hurricanes Hugo and Andrew. Similarly, organizations with prior disaster experience, including private firms, also appear to place increased emphasis on preparedness. For example, a recent study on changes in levels of preparedness among business firms following the 1994 Northridge earthquake found that among a wide range of businesses, those that sustained earthquake damage, those that were forced to shut down for some period following the earthquake, and those that suffered utility outages were subsequently more likely to increase their emphasis on preparedness (Dahlhamer and Reshaur, 1996). Focusing on the same event, Lindell and Perry (1998) reported that hazardous materials handlers stepped up their hazard assessment, mitigation, and preparedness actions after the earthquake. Interestingly, however, adoption of those measures could not be predicted by proximity to the earthquake's epicenter, onsite damage, offsite utility loss, or the cost of the damage businesses suffered.

At the community level, Drabek's extensive survey of the literature, completed in the mid-1980s, concluded that "[t]he greater the frequency that communities experience disasters, the more extensive will be their disaster planning efforts" (1986: 55). This finding has been replicated a number of times in research on local emergency preparedness agencies (Kartez and Lindell, 1990; Rogers and Sorensen, 1991), local preparedness networks (Gillespie and Streeter, 1987; Banerjee and Gillespie, 1994), and local emergency planning committees (LEPCs) (Lindell et al., 1996a; 1996b).

Higher levels of preparedness among social units with extensive disaster experience have sometimes been attributed to the existence of "disaster subcultures" (Moore, 1964; Anderson, 1965; Weller and Wenger, 1973). When informal groups, formal organizations, and communities have repeated experiences with disasters, such subcultures—consisting of beliefs about actions needed and a set of cultural defenses that constitute "a blueprint for residents' behavior before, during, and after impact" (Wenger, 1978:41)—may develop. Individuals with hurricane experience may decide for themselves what self-protective actions to take and when, instead of following warning advisories, for example. A case in point is South Carolina, where some residents evacuate when pre-hurricane winds reach a particular speed, rather than waiting until official warnings are issued. Over time, communities prone to seasonal flooding learn what to do when the water begins to rise and develop standard ways of coping.

Three factors are thought to foster the development of disaster subcultures. First, a community must experience repeated disaster impacts (Wenger, 1978). Second, those repeated impacts must result in significant damage. Third, advance knowledge of threats contributes to the emergence of subcultures. Repeated victimization in concert with serious repercussions can lead to prompt and effective response. For example, a chemical leak in Taft, Louisiana, in the early 1980s prompted 17,000 local residents to evacuate at 4:30 a.m. Their previous experiences with evacuations for hurricanes and other chemical threats provided an understanding of the danger and knowledge of how to respond (Phillips, 1992). In another example of subcultural adaptation to an ongoing threat, people who have extensive experience with earthquakes—particularly people from countries where earthquakes commonly cause structures to collapse completely—may opt to stay outdoors rather than use indoor shelters when an earthquake occurs, for fear of aftershocks. Community residents set up temporary living arrangements in parks, vacant lots, and other open areas, or they sleep in tents in their yards or campers parked outside their homes (Phillips, 1993; Bolin and Stanford, 1990).

However, studies on the impact of disaster subcultures do contain important qualifications. For example, Hannigan and Kueneman (1978) have suggested that as outside institutions increasingly take over tasks previously handled by households—such as preparedness and response—the influence of disaster subcultures may decline, since individuals and families may increasingly rely on organizations to carry out preparedness and response activities they once undertook themselves. Additionally, researchers also note that subcultural responses are not always adaptive. A generation ago, for example, Weller and Wenger (1973) hypothesized that experience with a particular disaster agent (e.g., seasonal flooding), rather than enhancing readiness, may instead engender a subculture of complacency, as households and communities learn to live with the hazard and accept losses. Conversely, disaster subcultures can develop that actually encourage risk-taking. The "hurricane parties" that have been documented in some hurricane-prone parts of the country are a case in point.

The idea that experience improves preparedness and response has not been supported across the board, either. Summarizing earlier research on the issue, for example, Drabek (1986: 107) observed that "[c]ertain types of disaster experiences appear to reinforce definitions of relative invulnerability," which may in turn lead people to discount real threats. Discussing the warning response literature, Perry and Mushkatel (1986) have noted that assumptions about the impact of experience that seem logical and consistent with behavior generally do not always apply empirically in disaster situations. As they pointed out (1986: 45):

> The theoretical arguments are strong and intuitively make it difficult to ignore the concept of experience in formulating explanations of warning belief and personal risk assessment. The empirical record is at best equivocal in describing the relationship between past experience and warning belief and risk assessment.

At the household level, Palm and her colleagues' study (Palm et al., 1990) on insurance found that earthquake experience—even direct earthquake damage to the household—was not a consistent predictor of the decision to adopt coverage. In fact, earthquake experience was related to higher levels of insurance coverage in only two of the four counties studied. Risk perception was the most consistent predictor of the decision to purchase earthquake insurance; respondents who considered their homes vulnerable to future earthquakes were more likely to obtain coverage than those who thought future events unlikely. Further, Palm and Hodgson's (1992) follow-up study on the impact of the Loma Prieta earthquake reported negligible effects of past experience on insurance purchase. Russell, Bourque, and Goltz (1995) also found experience to be an extremely weak predictor of household earthquake preparedness.

Similarly, at the community level, Kartez and Lindell (1987) cite a number of studies showing that "[l]ocalities often fail to improve their disaster plans even after major disasters," which again suggests that experience does not automatically lead to learning and further action. Another relevant study, which focused on the adoption of mitigation measures at the community level (Berke, Beatley, and Wilhite, 1989) found that disaster experience had no significant impact on community mitigation decisions.

Research also suggests that experience can have differential impacts on subsequent actions, depending both on contextual factors and on group characteristics. Reviewing a series of earlier studies on disaster warnings, Nigg (1987) observed that people with and without previous disaster experience may be influenced by different elements of the warning process. Those who have experienced a particular type of disaster tend to pay more attention to the content of a warning involving that particular disaster agent, while those lacking such experience may place greater emphasis on the credibility of the agency that issues the warning. In a study on responses to flood warnings, Perry and Mushkatel (1986) found that disaster experience was a significant predictor of warning belief for whites, but not for African-Americans or Mexican-Americans. Specifically, members of the latter group tended to believe the warning message regardless of whether they had previous disaster experience, while experience didn't affect African-Americans' propensity to believe. Those same researchers observed different relationships among experience, ethnicity, and warning belief in other disaster settings, suggesting that complex influences are at work that make generalizing about the impact of experience problematic. (Ethnicity and its impact on behavior in disaster situations is discussed in more detail in the section below.)

It is useful to note that some of the studies that did not find significant correlations between experience and preparedness did find significant relationships between experience and other variables that were in turn correlated with preparedness. Lindell and Whitney (1995), for example, concluded that disaster experience influences community support for preparedness, which in turn affects disaster readiness. In other words, there is evidence that experience can exert both a direct and an indirect influence on preparedness.

Perhaps part of the problem with attempting to assess the impact of experience on subsequent preparedness and response behavior lies in the varying and imprecise ways in which the concept has been measured. As operationalized by different researchers, "experience" can range from merely living in a community that went through some type of disaster, through directly sustaining severe and extensive losses in an event involving the same type of disaster agent such as repeated victimization in a flood or hurricane. Clearly, people may take away very different lessons from near-misses and minor disaster impacts than they would from disaster-related experiences that involve intense fear and large human and physical losses. Recent studies suggest that preparedness and response activities may be predicted better by the direct experience of adverse disaster consequences to the individual and his or her close associates (e.g., friends and relatives) than by measures of communitywide impact (Lindell and Prater, 2000). Similarly, Kartez and Lindell's (1990) survey of 370 local emergency management agencies suggests that the adoption of recommended preparedness practices is affected by experiences with specific types of disaster demands. Specifically, cities having more experience with response-generated demands, such as the need for coordination with other organizations and with the general public, developed more provisions to prepare for such demands.

The literature is also unclear on the generalizability of disaster experience. How much does experience with one type of hazard influence behavior with respect to others? Does exposure to a technological disaster agent such as a chemical explosion heighten concern about tornadoes? Should we expect it to? Additionally, existing research generally sheds little light on how individuals, organizations, and communities interpret their disaster experiences and what influences those interpretations. It has been recognized for more than 25 years that judgments about disaster vulnerability are influenced by various heuristics and biases (Slovic, Kunreuther, and White, 1974). As suggested above, certain types of experiences—for example, exposure to less severe events and mild impacts—may lead people to believe that disasters are not really anything to worry about and that preparing is not necessary. Alternatively, even experience with a very unusual and devastating event, such as a 500-year flood or a very large earthquake, may create a false sense of security if people reason that "We've had our big disaster, and so it won't happen again." Still another possibility is that a catastrophic disaster might produce a state of learned helplessness by leading people to conclude that it is useless to try to protect themselves against such events. If researchers begin doing a better job of measuring both disaster experience and how that experience is interpreted by victims, they should be able to learn a lot more about how experience influences subsequent preparedness- and response-related behaviors.

In sum, findings on the impact of disaster experience on subsequent preparedness and response behaviors are subject to many qualifications. While experience is clearly a factor that must be considered important in shaping preparedness and response over various social units and settings, beliefs and practices do not necessarily change as a result of actual events. Moreover, disaster experiences that contradict previously-held beliefs about disasters may simply be dismissed as anomalous (Quarantelli, 1984). Additionally, experience with one type of hazard may not carry over to others, particularly if those other hazard agents are perceived as very dissimilar. And it also appears that experience can teach the wrong lessons: communities that have extensive experience with a particular hazard may also shape their preparedness measures to deal with that one hazard, while ignoring other potential—and potentially more serious—kinds of threats.

SOCIOCULTURAL AND SOCIODEMOGRAPHIC FACTORS

In the past 25 years there has been a growing recognition of the ways in which a range of sociocultural and sociodemographic factors influence disaster preparedness and response activities. American communities are becoming more heterogeneous, and new studies indicate that population diversity affects emergency preparedness and response just as it influences social behavior generally. A detailed treatment of all the ways diversity matters and why it should be emphasized more in studies on preparedness and response would constitute a volume in itself—and one we believe is badly needed. In this section of the chapter, the most we will be able to do is begin discussing these differences in light of existing research and selectively report on research findings that illustrate their importance.

Ethnicity and Minority Status

Studies focusing specifically on racial and ethnic factors in disaster preparedness and response still are quite rare in the literature. However, a body of work is beginning to accumulate that underscores the importance of non-majority group membership in shaping the way segments of the population respond to hazards. For example, as Lindell and Perry (1992) note, racial and ethnic differences influence a wide range of perceptions and behaviors, including threat perception, concern about hazards, understanding of and belief in the science underlying hazard information, and attitudes towards the agencies disseminating information on preparedness.

Minority group members obtain their information on hazards from sources different from those used by members of the dominant majority group. The degree of credibility they assign to those information sources also varies, as does the nature and extent of their involvement with community organizations that can serve as a conduit for disaster-related information (Turner, Nigg, and Heller-Paz, 1986; Lindell and Perry, 1992). In a society like the U.S., which is characterized by a high and increasing degree of inequality (for just how profound those increases are see Wolff, 1996), it is not surprising that minority groups are also unequal with respect to their ability to protect themselves from disasters. Minority group members differ from the majority in their access to preparedness and other emergency-relevant information as well as in their responses to that information. These differences are attributable in part to language and income issues. More specifically:

> Minorities experience greater relative difficulties than whites because they have lower incomes and financial reserves, are more likely to be unemployed, less likely to have disaster insurance, and more likely to have problems in communicating with institutional providers of both information about disaster risks and post-impact relief (Lindell and Perry, 1992: 142).

Issues involving ethnicity and disaster vulnerability are complex. Recent research suggests that ethnic differentials exist not only between majority and minority groups but also among minority groups. For example, in their research on evacuation behavior in hurricanes Hugo and Andrew, Riad, Norris, and Ruback (1999) found that African-Americans were significantly less likely to evacuate than either whites or Latinos. Thus, to be meaningfully interpreted, research results need to be examined in light of exactly which minority groups are involved. It also appears that the nature of the differences observed between groups vary depending on the issue being studied—that is, on whether the focus in on preparedness, warning compliance, evacuation behavior, or recovery-related behavior (Perry, 1987).

With respect to warning response and evacuation, for example, researchers agree that compliance with warning instructions is influenced in a major way by three factors. The first of these is the family context in which the warning is received; unless and until the safety of all household members is accounted for, households are reluctant to heed evacuation directives. While no direct empirical examination of this proposition exists for different minority groups, it seems reasonable to assume that this generalization applies equally to all groups. At the same time, however, it is quite likely that the process of accounting for endangered kin may be more complex among some minority groups than among majority citizens, thus increasing the time required to decide upon and implement protective actions. We also know that household structure varies among ethnic groups, with American minorities tending to maintain more extensive and cohesive kinship networks that support extended families (Angel and Tienda, 1982; Ruggles, 1994). Hill and Hansen (1962) originally observed, and Perry, Lindell, and Greene (1982) subsequently confirmed in research on Mexican-Americans, that it is difficult for people living in extended family arrangements to account for family members and go through the warning confirmation process given the time and information constraints of the typical warning response period.

Second, research also indicates that the probability of heeding warning advisories is increased if warning recipients possess a personal adaptive plan or are aware of protective measures that could be undertaken to reduce danger. Here again, ethnic differences are likely to come into play in complex ways. Lindell and Perry (1992), for example, report that possession of adaptive information is positively correlated with warning compliance among both African-Americans and whites confronted with floods and hazardous materials incidents. Mexican-Americans who had adaptive information were also more likely to engage in protective measures, but often the action undertaken, while protective, was different from what authorities had recommended. Cautiously interpreting these data leads to the conclusion that possession of adaptive information is positively correlated with self-protective actions across all three ethnic groups, although understanding the specifics of the adaptive response may require consideration of other factors, such as family structure, message characteristics, and credibility attributions.

Third, warning compliance is also influenced by belief in the warning and by perceived risk. People who believe that a warning message is accurate and that it describes a real threat are more likely than others to comply with protective action recommendations. Although this relationship was empirically confirmed in a three-community study (Perry and Lindell, 1991) of African-Americans, whites, and Mexican-Americans, the study also found inter-ethnic variations in how that process developed. While there appeared to be no ethnic differences with respect to the positive relationship between warning belief, perceived risk, and compliance, the study did find that the three ethnic groups differed in the factors on which they relied to assess the risk and form a warning belief. Thus, while people generally are more likely to comply with warnings when they believe the risk is high, the process of determining what constitutes high risk could vary substantially between members of different ethnic groups.

It also is highly likely that ethnicity exerts its influence on relevant disaster behaviors via indirect or interactive effects with other variables. At least three such variables have appeared in the disaster literature in connection with ethnicity: socioeconomic status, perceptions of the credibility of authorities, and the psychological construct known as locus of control.

Looking first at socioeconomic factors, the available research indicates that minorities find it harder to cope with disaster because they tend to have less wealth and lower incomes, and also because they are more likely than whites to experience problems in communicating with authorities (Dacy and Kunreuther, 1969; Baumann and Sims, 1978). However, ethnicity and income are also associated with differences in other factors that affect the ability to cope in disaster situations, such as education and access to social support networks (Riad, Norris, and Ruback, 1999). Because these different influences are often so strongly correlated in the U.S., it is difficult to disaggregate their effects without conducting studies using large samples and multivariate analytic techniques. Unfortunately, to date there have been too few systematic empirical studies to reach solid conclusions on the independent influence of these factors.

With respect to the credibility and believability of hazard-related messages, a variety of studies document that minority citizens are less likely than the majority population to perceive majority group authorities as credible information sources (McLuckie, 1970; Staples, 1976). This is significant, since perceptions of credibility are linked with warning belief and perceptions of personal risk. In studying African-Americans, Mexican-Americans, and whites, for example, Perry and Lindell (1991) found that when asked to identify credible sources of hazard information, white respondents were more likely to identify public authorities (police and fire departments) and mass media. While blacks also found authorities credible, they cited social network sources (relatives, neighbors, and friends) as credible sources more often than whites, and they rarely relied on the media. The distinctive pattern of Mexican-Americans was to place highest confidence in the credibility of social networks. These findings must be interpreted cautiously, however, because only two communities were studied, and all of the minority group members in the study had comparatively low incomes, once again confounding ethnicity and income factors.

Finally, some disaster studies suggest that locus of control is correlated with warning compliance and the adoption of protective measures. Sims and Baumann (1972) reported that internal locus of control—or the extent to which individuals believe they can control events in their lives—is positively correlated with protective responses to tornadoes. Three other studies also suggest that ethnicity is connected with locus of control, but they have the disadvantage of dealing with only three ethnic groups. Ives and Furuseth (1980: 14) found that "a significant subgroup of Blacks . . . view flooding as an uncontrollable natural event and are less confident in their ability to deal with the hazard." Similarly, Turner, Nigg, Heller-Paz, and Young (1981: 3) report that African-Americans and Mexican-Americans "were more fatalistic about earthquake danger and skeptical about science and the predictability of earthquakes" than whites. Finally, Perry, Lindell, and Greene (1982) reported that Mexican-Americans (in California) were more likely to possess an external locus of control and were less likely to have developed family flood emergency plans than African-Americans and whites.

Response-related behaviors other than warning response and evacuation are also affected by ethnicity. With respect to post-disaster shelter-seeking, Perry and Mushkatel (1986) found that, although the African-Americans, whites, and Mexican-Americans in the three disasters they studied all tended to seek shelter with family members or friends, there were also ways in which the three groups differed. For example, in the large city they studied, African-Americans were more likely than the other two groups to use public shelters, and in the non-urban area, Mexican-Americans were less likely to do so. As noted above in our discussion of disaster subcultures, several studies on the public response to earthquakes suggest that Latino community residents—particularly recent immigrants from Mexico and Central America—prefer outdoor sheltering over the use of publicly-designated shelters and other forms of indoor sheltering (Bolin and Stanford, 1990). In her study of sheltering following the Loma Prieta earthquake, Phillips (1993) attributed this pattern to residents' prior experiences with damaging earthquakes in their native countries, their desire to stay close to their homes to keep an eye on their property, agencies' lack of planning to provide shelter to a culturally diverse population, and the general failure to include the Latino community in pre-disaster planning.

Some ethnic group members in the United States may also have immigration-related concerns that influence the ways in which they respond when a disaster occurs (Bolin, 1998). They may, for example, avoid public shelters and other services for fear of being discovered by the Immigration and Naturalization Service and deported—a fear that is justified, since that is what happened to Hispanics and Haitians following Hurricane Andrew (Phillips, Garza, and Neal, 1994). Others may fail to apply for services to which they are entitled because they believe incorrectly that they are not eligible, or because information on programs has not reached them. In the current legal and social climate, many minority group members and immigrants, including those who are U.S. citizens, justifiably feel singled out for special scrutiny by authorities. These kinds of feelings influence their behavior during disasters, just as they do in other situations. Since an increasing number of government services of all kinds are being denied to both undocumented and legal immigrants, confusion and concern about program eligibility are likely to grow within minority communities. Those concerns will in turn influence sheltering and other patterns of post-disaster service utilization among affected groups.

This book deals with preparedness and response issues, but we should also note that ethnicity and minority status are factors that need to be taken into account throughout the entire hazard cycle. Perry and Mushkatel (1984), for example, have pointed to the need to consider the needs of minority communities when decisions are made to relocate communities or neighborhoods to reduce future losses. Studies by Bolin (1982) and Bolin and Bolton (1986) have found that experiences during the recovery period are structured by race and ethnicity, producing slower disaster recovery among minority-group members. Recent laws and changes in entitlements will undoubtedly make some groups' struggles to recover following disasters even more difficult than before. For example, following the 1994 Northridge earthquake, disaster-relief legislation explicitly closed off all but emergency forms of assistance to undocumented residents in the impact area, making them ineligible for the longer-term types of aid that are designed to help families recover. Recent welfare legislation may also have adverse effects on minority groups and the poor (see discussion below).

Language

For non-English speakers, language can constitute a barrier to involvement in the emergency planning process and can also influence response behavior in many ways. These range from limitations on the ability of non-English speakers to hear and comprehend warning messages to problems with access to information on options for sheltering and other services. Since language differences tend to restrict people's community participation to within their own language groups, non-English-speakers may lack access to disaster-related information and programs that are available to the rest of the population. More generally, language differences may also cut people off from the planning process itself unless special steps are taken to involve non-English speakers.

Disaster preparedness and response can be further complicated if appropriate translators are not immediately or conveniently available; if they incorrectly translate information; if there are dialect variations within the same language group that create translation problems; or if messages are unwittingly distorted or made ambiguous in the process of translation (Perry, Lindell, and Greene, 1981). During the response period, language differences may lead warning recipients to delay the initiation of recommended protective actions, take inappropriate action, or simply not act at all because of failure to understand what the warning message was trying to convey. In sum, language differences are a key factor in understanding pre- and post-impact behavior.

The Saragosa, Texas, tornado of 1987 is perhaps the most glaring recent example of the failure to reach non-English-speaking community residents with appropriate warning information and of why it is so important to take language differences into account. At the time of the tornado, Saragosa was a small, unincorporated town of about 400 people, most of whom worked on nearby farms, ranches, and in service establishments. Virtually all the families were of Mexican descent, and the majority of residents spoke only Spanish. The large tornado that struck the town in May of 1987 killed 30 and injured 120. The residents of the community had received no official warning that the tornado was coming—or more accurately, no official warning they could understand that was broadcast through media they typically used. The local media that disseminated the warnings handled the Spanish-language warning messages badly. The translated warning messages, which were improvised on the spot, did not convey how severe the danger was, and the Spanish-language warnings were disseminated later than the ones in English. The local Spanish-language cable television channel, which most people watched, did not broadcast the warning. The residents of Saragosa were thus put at higher risk, with tragic results, because the warning system failed to consider their needs. One of the key conclusions of the National Academy of Sciences report on that tornado (Aguirre et al., 1991: 2) was that "warnings, to be effective, require either a common shared culture or adaptation of the warning system to multicultural social contexts. In Saragosa neither requirement was satisfied." (For additional information on this disaster see also Aguirre, 1988.)

Although research in this area still is very sparse, the suggestions for improving emergency management for populations containing large concentrations of non-English speakers tend to focus on improving outreach in the area of emergency planning. For example, Perry and Greene (1982) have argued that those responsible for preparing the community need to identify groups of non-English speakers in a systematic fashion as part of the process of planning how to disseminate emergency-relevant information. In cases where a single non-English-speaking language group is represented—for example, Spanish-speakers in many parts of the Southwest—it is feasible to prepare both written and verbally presented materials (including warning messages) in the language and then transmit messages through channels specifically targeted to that group of non-English speakers. In other cases, written translation is more problematic. Examples include communities where there are multiple language groups, languages that do not have written forms, and target populations that are not literate. In such cases, strategies involving direct contact, such as neighborhood meetings, and approaches that employ visual images rather than only written text may be useful. One important strategy for reaching underserved populations is to expand efforts to involve members of minority communities in key emergency management posts, because ethnic minority groups are probably more likely to accept and adopt measures that are presented by someone from a common background (Perry and Mushkatel, 1986).

Clearly it is also important to involve non-English mass media outlets in reaching non-English-speaking populations with disaster preparedness messages and providing needed information during the response period. Disaster-related information will not reach its intended audiences unless it is presented in an understandable way, through the same channels non-English speakers typically use for obtaining information of other kinds.

In the section above, we noted that current policies toward immigrants are likely to affect minority group vulnerability in future disasters—particularly those groups most affected by efforts to deny services to undocumented persons and non-U.S. citizens. In this same vein, the growing emphasis on "English-only" programs and on English as the "official language" for governmental business could seriously complicate efforts to reach non-English speakers with preparedness information and to provide services to them in disaster situations. Such changes should be monitored by researchers, and their impact on emergency preparedness and response activities should be documented.

Social Bonds

Social attachments and relationships are key predictors of the preparedness and response behaviors undertaken by different social units. Strong and extensive social bonds generally have a positive effect on emergency response-related behaviors. Social connectedness, measured in various ways, has been shown to foster adaptive behavior during both the pre- and post-disaster periods. For example, with respect to hazard education programs for community residents, Lindell and Perry argue that "residents' access to such hazard awareness programs will be a function of their community involvement" (1992: 140). Above we pointed out the importance of community attachments such as home ownership and of the parent-child bond in stimulating household disaster preparedness. Turner, Nigg, and Heller-Paz (1986) found that community bondedness—defined as neighborhood tenure, identification of the neighborhood as home, participation in community organizations, and the presence of friends and relatives nearby—was significantly correlated with preparedness for earthquakes. Strong bonds were in turn positively related to income and to the presence of children in the home.

As noted above, one of the most durable findings in the household warning response and evacuation literature is that household context—for example, whether household members are together when they receive the warning message and whether they are able to evacuate as a unit—shapes responses (Perry, 1982; Drabek, 1983a; Perry and Greene, 1983). Family ties and other forms of social involvement are also factors in the receipt of warning messages, in part because people use their social networks to confirm officially-disseminated warning messages. And when people do leave their homes to seek emergency shelter or temporary housing, social ties are a key factor determining where they go; people who are able to do so prefer to stay with relatives, neighbors, and friends.

The nature and extensiveness of social bonds are related in part to ethnicity. Ethnic groups differ, for example, in the extensiveness and intensity of their kinship relationships and in levels of community participation. Lindell and Perry (1992) cite various studies suggesting that the characteristics of different groups' social bonds—e.g., the propensity to be involved in extended as opposed to nuclear family forms—affect their response to hazards. Since members of minority groups participate differentially in different types of community activities—such as voluntary associations, clubs, and political and school-centered activities—they are likely to have access to different types and amounts of hazard-related information. This can in turn influence the preparedness and response actions they take. To date, however, not much systematic research exists on how those influences actually operate.

Moving from households to the organizational and interorganizational levels, sociological research documents the many ways in which ties among organizations are important both for constituent organizations and for the networks to which they belong. For example, interorganizational linkages are major sources of information transfer and of access to new ideas and innovative practices; they make it easier for organizations to act in concert and mobilize their members; they serve as the basis for various kinds of exchanges, such as the exchange of political favors and other forms of assistance; and an organization's position in interorganizational networks is a key source of both perceived and actual influence. (The literature on interorganizational relations is quite large, but for good discussions on the significance and impact of networks see Boje and Whetten, 1981; Marsden and Lin, 1982; Knoke, 1990.)

To the extent patterns of contact and coordination among organizations have been systematically studied in the disaster literature, research has supported these more general sociological findings. Both older and more recent research and both qualitative and quantitative studies document the significant impact various kinds of multi-organizational and organization-environment relationships have on planning and response. The literature stresses the importance of interorganizational and multi-organizational (as opposed to single-organization-focused) preparedness; the problems that can occur when organizations plan in isolation; and the impact interorganizational ties (communication, resource exchanges, and coordination) exert on planning and response. Gillespie et al. (1992) found that various forms of pre-disaster networking among community organizations fostered community response effectiveness. In a study of 12 communities, Nigg (1987) found that successful emergency management programs were those in which directors actively sought to develop and maintain interorganizational networks. Drabek's (1990) discussion on the strategies employed by effective emergency managers stresses the efforts they made to establish and maintain ties with other organizations, including holding joint disaster exercises with other organizations, conducting community seminars, holding regular committee meetings, distributing printed communications such as newsletters, assigning individuals to act as liaisons with other departments and organizations, and appointing advisory committees. In other words, in large measure these leaders were effective because their efforts were focused outward, into the larger community, rather than inward.

Building local emergency response capacity thus appears to involve the ability to pursue a variety of bridging and boundary-spanning activities, such as maintaining frequent interdepartmental and interorganizational communications; establishing councils, boards, and mutual aid networks representing key organizational actors in the community; organizing joint activities such as communitywide disaster exercises; and attempting to make emergency operations centers vehicles for interorganizational coordination.

Income Inequality and Economic Resources

The impact of resources of various kinds on preparedness and response activities is evident both at the household level and among other social units. The importance of financial resources for household preparedness is apparent in the positive relationship that generally exists between household income and adoption of preparedness measures. Higher-income households are more likely to be insured against hazards, in part because they are more likely to own their own homes. Home ownership is, of course, another indicator of financial well-being, and, controlling for other relevant factors, property owners are more likely to prepare for disasters than renters. Renters typically also lack the ability to undertake structural mitigation measures that can protect them in the event of a disaster. When a disaster does occur, they are also dependent on their landlords to make necessary repairs and upgrades.

Since income is positively related to access to better and safer housing, low-income households are at greater risk from many hazards. Hurricane Hugo and the Loma Prieta earthquake had a disproportionately severe impact on poor and minority residents because these were the groups most likely to live in overcrowded, substandard, and easily-damaged housing (Simile, 1995). Older, unreinforced masonry buildings, which are prone to collapse in earthquakes, constitute an important source of affordable housing for lower-income residents of earthquake-prone cities like San Francisco and Los Angeles. In the 1989 Loma Prieta earthquake, the cities of San Francisco and Oakland lost a significant proportion of their low-income housing due to earthquake damage; single-room-occupancy hotels and homeless shelters were particularly hard-hit. Following that earthquake, FEMA was severely criticized for disaster assistance policies that unfairly discriminated against low-income households, members of the homeless population, and people in transient living situations (U.S. General Accounting Office, 1991). Mobile homes, another source of housing used primarily by low-income people, are also highly susceptible to disaster damage. In 1994 nearly 40 percent of all tornado fatalities occurred in mobile homes (U.S. Department of Commerce, 1995). Of course, higher-income people do lose their lives, homes, and livelihoods in disasters, but other things being equal, lower-income people are disproportionately exposed to the risk of being killed, injured, or displaced by disasters.

Vulnerability to technological hazards is also structured by economic inequality, as well as by race and ethnicity. A body of work has begun to develop showing that poor, minority, and less politically-powerful communities are disproportionately exposed to the hazards associated with toxic waste sites and other noxious facilities (Bullard, 1990, 1994; Rosen, 1994; Krieg, 1995, 1998). Even when income is held constant, communities with large minority populations are more likely to be exposed to such hazards, and government agencies also act more slowly and spend less to ameliorate toxic hazards affecting minority communities (Bullard, 1994). Poor and minority communities have begun to mobilize to press for the remediation of imposed environmental hazards. The social construction of an environmental justice frame (Capek, 1993) that places those risks in a broader sociopolitical context and defines protection from toxic substances and facilities as an inalienable right has been a key element in that mobilization.

Critics of the environmental racism/environmental justice literature argue that correlations that may exist between race and exposure to environmental hazards do not necessarily imply causation, and some studies have found no significant linkage between race and toxic threats (Anderton et al., 1994; Anderton, Oakes, and Egan, 1997). Other recent studies find that there is a link, but that the relationship is complex, and that working class communities, as opposed to the very poor or the well-off, are most at risk (Been and Gupta, 1997). Another approach, suggested by Krieg (1998), argues that higher levels of exposure to environmental toxins are one consequence of community dependence on low-wage polluting industries, a pattern that is often, but not always, associated with race.

The impact of monetary resources is also apparent in research on the kinds of measures households are most likely to adopt to prepare for disasters. For example, in studies of earthquake preparedness (see Edwards, 1993) relatively inexpensive, easy strategies—such as having a battery-operated radio, having a working flashlight, and storing water—were far more prevalent than more expensive or time-consuming strategies.

It is also likely that financial resources are significant for preparedness at the organizational and interorganizational levels, although here the linkages are less clear. Organizational size is one measure of resources, and the literature generally finds a positive relationship between size and preparedness. For example, what little research exists suggests that larger businesses are more likely to prepare for disasters than smaller ones (Drabek, 1994; Dahlhamer and Reshaur, 1996; Dahlhamer and D'Souza, 1997). Intuitively we would expect that budgetary resources would be important predictors of preparedness among public agencies, but empirical evidence for that relationship is virtually non-existent. In the private sector, we can also hypothesize that, other things being equal, better-off companies will be more likely to have the resources to apply to preparedness than those that are in a precarious financial position. Again, while some research exists supporting this notion (Dahlhamer and Reshaur, 1996), the relationship has yet to be explored in depth.

Similarly, researchers cannot yet say with confidence whether better-off communities are more likely to engage in planning than those that are less affluent. The work that has been done linking resources and preparedness tends to measure resources at the organizational level (e.g., Lindell and Meier, 1994) rather than taking the broader community context into account. However, research on Hurricane Andrew does suggest ways in which poorer communities are at a disadvantage when disaster strikes. Comparing a better off, primarily white community with a significantly worse off, predominantly African-American community, both of which were struck by the hurricane, Dash, Peacock, and Morrow (1997) show that housing, job, business, and tax revenue losses were proportionately greater in the minority community. At the same time, the poorer community was less able than its more affluent counterpart to manage recovery efforts in the post-disaster period because of major personnel and organizational problems and the fact that "[its] normal disadvantage in Dade [County]'s political and economic structures was further crippled by its lack of experienced administrators and staff as it attempted to deal with the complex problems of recovery" (Dash, Peacock, and Morrow, 1997: 217). This small minority community, which had more problems to begin with due to the vulnerability of its building stock, was ill equipped to face the complex demands produced by a massive disaster.

It seems reasonable to assume that community fiscal well-being is a necessary (but likely not sufficient) condition for effective community disaster management. Given the generally low priority assigned to disaster issues among most U.S. communities, it is not difficult to hypothesize that, in communities experiencing budget problems, disaster-related programs would be among the first to go, that existing funds would be allocated to problems perceived as more pressing, or that improvements in disaster readiness would be defined as too expensive. We have anecdotal evidence of these kinds of patterns but little in the way of systematic research. Additionally, while lack of funds could translate into lack of preparedness and response capability, more may not always mean better; prosperous communities may be no more willing than their less-well-off counterparts to invest in safety, or they may spend on the wrong things. At this point, however, there is not enough research on these kinds of issues to say much with confidence on the relationship between community-level economic and fiscal factors and preparedness and response.

Worldwide, it is clear that higher levels of affluence are associated with lower levels of disaster vulnerability, particularly in terms of lives lost in disasters; the damages wrought by disasters in the Third World far exceed those experienced in developed countries. While in absolute terms the monetary losses from disasters in advanced capitalist countries like the U.S. and Japan can be enormous (an estimated $40 billion for the 1994 Northridge earthquake and $120 billion for the 1995 Kobe earthquake, for example), compared with gross domestic product and the overall investment in the built environment such figures are not large. And if losses of this magnitude are judged unacceptable, then developed societies have the economic resources to pay for higher levels of safety—although whether they choose to is a separate issue. Countries that cannot feed their people might understandably view spending on disaster programs as a luxury, and in many parts of the world disaster vulnerability increasingly pales in the face of more pressing problems such as war, economic dislocation, genocide, and forced migration. Many societies today exist in a more or less permanent state of crisis that both constrains the ability to plan for disasters and at the same time makes them more likely. We will return to these themes in Chapter Seven, which places disaster preparedness and response in the broader context of sustainability.

As we have noted elsewhere in this volume, social change continually affects both disaster vulnerability and the ability of social units to prepare and respond when disaster strikes. Non-whites tend disproportionately to be poor, and the proportion of ethnic minorities in the U.S. population is increasing. Changes that adversely affect the economic fortunes of groups within society will also have an impact on their safety in disaster situations and in their ability to prepare, respond, and recover. While most people probably do not see much connection between legislation like the 1996 welfare law and the ways in which disasters may affect our society in the future, it is quite likely that such a connection exists. If economic inequality continues to increase, so will the problems of poor people facing disasters. If certain types of income support are denied to immigrant groups, that may well lower the ability of immigrants to protect themselves against disasters. If services are not available to immigrants when they experience disaster-related losses, that will almost certainly affect their ability to recover. Disasters and their effects must be seen in the broader societal context, and the impact of these kinds of changes on vulnerability is an important area for future research.

Gender

Along with race and social class, gender is a major stratifying force in society. Since institutional patterns are invariably gendered (Acker, 1990, 1992), it would be surprising indeed if the impact of gender differences were not felt in disaster situations. However, with the exception of a handful of scattered references in the classic literature on disasters and a few recent works that look at the topic in more detail, gender has generally not been a focus in the literature on disaster preparedness and response. Gender issues, particularly as they intersect with class and race, are only beginning to be considered in disaster preparedness and response research (Morrow and Enarson, 1994; Fothergill, 1996, 1998; Scanlon, 1997; Enarson and Morrow, 1998). Much of the best research on gender and disasters has been conducted in countries other than the U.S., particularly developing countries, and has focused on the ways in which globalization and other trends in the political economy of the world system reinforce the gendered division of labor, frequently increasing women's disaster vulnerability (Blaikie et al., 1994; Enarson, 1998; Fordham, 1998).

In their recent edited volume on gender and disasters, Enarson and Morrow (1998: 4) discuss the importance of adopting a gendered perspective when studying disaster-related phenomena:

> The social experience of disaster affirms, reflects, disrupts, and otherwise engages gendered social relationships, practices, and institutions. Disasters unfold in these highly gendered social systems. Disaster management is correspondingly engendered, shaping the environmental decisions we make and contingencies we fail to plan for, the dynamics of our disaster-management organizations and relief operations, the disaster-responding household and emergent response groups, the decisionmakers we choose and the heroes we create.

As this passage suggests, hazard-related gender issues span a range of role-related, economic, familial, occupational, organizational, and political concerns (Schroeder, 1987; Phillips, 1990; Neal and Phillips, 1990; Morrow and Enarson, 1994; Enarson and Morrow, 1998). In a review essay that summarizes and synthesizes research findings on gender and hazards, Fothergill (1996; see also Fothergill, 1998) cites a number of studies suggesting the various ways in which gender is relevant for our understanding of hazard vulnerability, hazard- and disaster-related behavior, and disaster impacts and recovery. According to Fothergill, gender plays a key role throughout the hazard cycle, explaining differences in exposure to risks, risk perception, preparedness and response behaviors, vulnerability to physical disaster impacts and to the psychosocial consequences of disasters, and participation in response- and recovery-related activities.

Women are often more vulnerable to hazards, both because of their role-related caregiving responsibilities and because of their greater tendency to be living in poverty (Fothergill, 1996, 1998). Gender differences in disaster-related mortality and morbidity have also been documented. Studies indicate that some disaster agents and events have killed and injured more men, while others have disproportionately affected women. Such patterns are traceable not only to role-related behaviors but also to male/female power inequities. For example, women's roles in caring for children involve them in a daily round of activity that may make them more vulnerable to certain kinds of disaster agents, such as earthquakes and associated building collapses, while protecting them from others, such as flash floods and lightning (Fothergill, 1996, 1998). A number of studies conducted in developing countries have found that women and girls are more susceptible to the effects of famine and drought, both because of the greater power and the privileges that accrue to males in those societies and because of broader political-economic forces. When food is scarce, it is usually the women who receive less adequate rations, resulting in higher rates of female mortality. (For a more detailed discussion of the literature on gender and famine vulnerability and of the relevance of this literature to gendered research on disasters see Bolin, Jackson, and Crist, 1998.) Gender may also be linked to the risk of injury when an earthquake strikes. Some findings suggest, for example, that women may be at risk during earthquake shaking as they move to protect and comfort their children, while men's vulnerability may stem from their greater tendency to try to run out of buildings (Bourque, Russell, and Goltz, 1993).

Women have also been found to perceive risks differently from men and generally to be more risk-averse (Cutter et al., 1992). In a three-community study, Hamilton (1985) found concern with toxic hazards to be high among women, as well as among younger respondents and those with children under age 18. In another study focusing on technological and environmental health risks, Flynn, Slovic, and Mertz (1994) found that risk perceptions among white males differed both from those of white females and from those of non-white males and female, in that white males were markedly less likely to see a range of hazards as risky than were members of those other groups. White males' views on hazards appear to be based on their higher levels of trust in institutions and their greater willingness to see people as responsible for making their own choices about the risks they face.

Although there is little research in this area, women's involvement in preparedness activities may also differ from men's. It has been found, for example, that emergent groups that form to combat health risks from hazardous materials dumps and other environmental problems typically consist primarily of women (Neal and Phillips, 1990). Fothergill (1996: 38) observes that "women become active in these groups through friendship networks and because disasters pose a threat to the home and the community; thus, women's membership is seen as an extension of their traditional domestic roles and responsibilities." Lindell and Prater (2000), replicating previous studies, found that women had higher levels of seismic risk perception than men. Women also reported higher levels of hazard intrusiveness—that is, they thought and talked more about earthquakes than men. Nevertheless, women carried out fewer mitigation and preparedness actions, which may be traceable to their generally lower incomes.

Men and women also appear to respond differently in both the warning and impact phases of disasters. For example, women are probably more likely than men to give credence to warnings and to want to evacuate in the face of an impending threat (Drabek, 1969; Beady and Bolin, 1986; Riad, Norris, and Ruback, 1999). Similarly, some response-related behaviors also appear to be gender-based. In fire situations, for example, women are more likely to warn others, while men are more likely to try to fight the fire (Wood, 1980). Other evidence of role carryover in disaster situations can be seen in women's greater propensity to become involved in food preparation and other supportive activities during the response period, as opposed to the active rescue roles that disproportionately attract men (e.g., Wenger and James, 1994; but see also O'Brien and Mileti, 1992, who found no relationship between gender and participation in response activities).

The literature also suggests that women's vulnerability extends into the post-disaster period. Women may experience higher levels of emotional stress than men following disasters, and those impacts may be of longer duration (see, for example, Green et al., 1991; Bolin, 1994). These emotional strains may stem from a variety of sources, including the expansion in women's caregiving roles and social-support activities following disaster, their lack of financial resources with which to recover, or their unequal access to recovery assistance (Enarson and Morrow, 1997, 1998). Women may also be more vulnerable to criminal activity, including both domestic violence and disaster damage repair scams undertaken by unscrupulous "contractors" (Peacock, Morrow, and Gladwin, 1997; Enarson and Morrow, 1998).

This emerging awareness of the significance of gender also signals the need for a gendered approach to the study of the organizations that are involved in emergency preparedness and response (Enarson, 1998; Tierney, 1998). The field of emergency management is an excellent example of what Acker (1990, 1992) would term a male-gendered occupation. The emergency management function is most typically housed in civil defense agencies and fire and police departments—organizations that until quite recently have been composed almost exclusively of men. Like other public safety organizations, emergency management agencies may stress personal qualities, action styles, and modes of organization that are more compatible with men's perceived gender roles than with women's, such as risk-taking, aggressiveness, an emphasis on rapid decision making and action, and a preference for hierarchical forms of organization. Although less so than in the past, prior involvement in the military or the uniformed public services is still considered an important qualification for a job in emergency management. The number of female emergency managers has increased considerably in recent years, but the job likely remains one for which men are perceived as better suited than women. The topic of gender in crisis-related organizations is a topic that is ripe for future research. The few studies that have focused on the subject (see, for example, Chetkovich, 1997, on the urban fire service) suggest that hostility toward women and toward attitudes and behaviors that are defined as feminine persist in emergency-oriented organizations in spite of changes in the gender composition of those agencies.

The gendered nature of the emergency management profession undoubtedly has affected hazard management activities in many ways.

However, we are unaware of any systematic empirical research that attempts to explore these kinds of influences. Similarly, while more women are working in the emergency management field, no systematic studies have been conducted, either to explore how women have adapted in these largely male-dominated organizations or to asses the impact of their entry into the field.

Although we do argue here that a gendered perspective will contribute a great deal to our understanding of hazard-related behavior, we agree with Bolin, Jackson, and Crist (1998) that it is important to avoid overly simplistic and essentialist notions about gender and disasters. As other recent work in the social sciences shows (see, for example, Baca Zinn and Dill, 1994; Andersen and Collins, 1998), gender issues have to be seen in the context of the racial, social class, and power inequalities that also structure social, organizational, and interpersonal behavior. Research should not be conducted in such a way that "women and men are reduced to separate universal constructs undifferentiated by class or culture or experience" (Bolin, Jackson, and Crist, 1998: 35). Nor should our analyses assume that social action consists merely of the playing out of gender-role scripts. If this overly-structured approach is inadequate to understanding everyday social life, it is even less appropriate for the study of disaster-related social activity, since disasters by their very nature render problematic many taken-for-granted aspects of daily life and encourage improvisation and emergence. The occurrence of a disaster may propel women who were previously concerned primarily with their domestic duties into activism in the public arena. Traditional gender-based divisions may break down in the face of disaster-induced pressures, and the demand for labor and services in the aftermath of a disaster may create employment opportunities for women (Enarson, 1998; Enarson and Morrow, 1998). Just as poor people and members of ethnic minorities have mobilized following disasters to press for recognition of their interests, disasters also create the possibility for gender-based mobilization, as was documented in the aftermath of Hurricane Andrew, when women formed the organization called Women Will Rebuild in response to the marginalization of their concerns by post-hurricane recovery groups such as We Will Rebuild (Enarson and Morrow, 1998). Thus, even as pre-disaster patterns of difference and inequality—including those linked to gender—carry over into disaster situations, disasters also create opportunities for innovation, non-traditional solutions to problems, and challenges to the existing social order.

Other Socioeconomic and Sociocultural Factors

Personal attributes such as age and physical capacity are also likely to play a role in preparedness and response. But, again, this is an area in which little research exists. With respect to age, contrary sets of research findings suggest on the one hand that elderly persons experience deprivation relative to their younger counterparts (Friedsam, 1962; Bolin and Klenow, 1988) and on the other that, by virtue of life experiences or social support, older people are able to avoid the negative effects of disaster (Huerta and Horton, 1978; Murrell and Norris, 1984; Melick and Logue, 1985; Kaniasty and Norris, 1993; Norris, 1992). Both conceptual and empirical studies (Durkin, Aroni, and Coulson, 1984; Tierney, Petak, and Hahn, 1988; Johnson, Johnston, and Peters, 1989; Gulaid, Sacks, and Sattin, 1989; Vogt, 1991) provide support for the idea that elders suffer more injuries and loss of life in disasters than do younger people. There are two plausible reasons for these patterns. First, the physical disabilities that are correlated with age likely put elderly people at a disadvantage in emergency situations, particularly if rapid action or physical exertion is required. Second, elderly people—particularly those with limited financial resources—may also be more likely to reside in disaster-vulnerable structures. In the 1995 Kobe earthquake, for example, mortality was strongly correlated with age, because older people tended to live in traditionally-constructed houses that were more likely to collapse. Further, the elderly were more likely to live in the densely-populated, lower-income sections of the city that burned following the earthquake (Tierney and Goltz, 1997).

Much of the research showing that elderly persons are more resilient to disasters derives from psychological research on mental and physical health that often finds minimal differences based on age (Melick and Logue, 1985). In some studies, older adults have been found to have better mental health than their younger counterparts (Murrell and Norris, 1984). Some research suggests that disaster severity may be related to the fate of elderly persons, with heavy devastation increasing problems for the aged (Phifer and Norris, 1989).

Only minimal research exists on persons with disabilities in disaster situations (but see Tierney, Petak, and Hahn, 1988; Rahimi and Azevedo, 1993). Since the time of the first assessment, however, agencies such as FEMA and the Red Cross have become more aware of disaster victims with disabilities. As the U.S. population continues to age, it will become increasingly important to consider elderly people, particularly those with disabilities, in preparedness and response planning.

Social Inequality, Diversity, and Disasters

Since the mid-1990s new work has begun to appear that explicitly focuses on the role social inequality and population diversity play in hazard vulnerability. This emerging disaster paradigm views disasters as the product of both physical forces and social-structural factors that combine to place individuals and groups at risk. In their influential book *At Risk*, for example, Blaikie et al. (1994: 3) argue that:

> [t]he crucial point about understanding why disasters occur is that it is not only natural events that cause them. They are also the product of the social, political, and economic environment (as distinct from the natural environment) because of the way it structures the lives of different groups of people.

This view has much in common with the natural hazards paradigm proposed much earlier by White and his colleagues, which we discussed in Chapter One (White and Haas, 1975; Burton, Kates, and White, 1978). However, it modifies and extends that approach by arguing that vulnerability to disasters is shaped in large measure by socioeconomic factors, including social class, race and ethnicity, gender, age, and rights over property. This is the case because such factors influence the access people have to the resources they require in order to be safe and secure, including not only monetary resources but also the information, social network ties, and sources of support they need in order to avoid disaster impacts or to cope and recover if they experience a disaster.

At various points in this chapter, we have discussed research that documents the ways in which socioeconomic and sociocultural factors influence exposure to hazards and strategies for managing them. Financial resources help determine which self-protective measures people adopt as well as affecting their access to various forms of post-disaster aid. Financial resources buy higher levels of safety; poor people tend to live in poor-quality housing that is vulnerable to disaster damage. Language, race, ethnicity, and social networks influence both how people perceive disaster warnings and what they do in response to those warnings. Larger and more profitable businesses are better-prepared for disasters than their smaller counterparts and more likely to recover when a disaster does occur.

This approach to explaining disaster vulnerability aligns theorizing about disasters more closely with other research on the ways in which social factors shape life chances and life experiences. It has long been understood that class, race, gender, and ethnicity are related to many other forms of vulnerability, including differences in rates of physical and mental illness (Kessler and Neighbors, 1986; Kessler, Turner, and House, 1989; McLeod and Kessler, 1990); vulnerability to violent crime (Blau and Blau, 1982; Shihadeh and Steffensmeier, 1994; Hagan and Peterson, 1995; Martinez, 1996); and mortality and life expectancy (Shrestha, 1997). It should come as no surprise, then, that such factors also play a role in exposure to disasters and their effects. One marked change in the research literature since the last assessment involves the extent to which the role that social inequities play in disaster victimization is being explicitly acknowledged.

The literature on the social-structural factors associated with disaster vulnerability has grown significantly in the past 25 years, but much remains to be learned. Future research needs to explain how cultural diversity and social inequality influence preparedness and response activities. Besides being important from a theoretical perspective, research of this type has direct implications for disaster policy and service delivery because, when disaster-related programs neglect to consider the needs of an increasingly diverse population, a host of problems can develop. For example, mass-feeding operations may fail to take into account the dietary preferences of some population groups, and providers of emergency shelter and temporary housing may overlook the needs of chronically ill and elderly victims. An able-bodied military may prepare a tent city without considering persons with physical disabilities (Neal and Phillips, 1995). Lack of awareness and prejudice may mean that communities or organizations deny people with AIDS access to shelters. Aid programs that effectively reach majority-group populations may miss members of minority groups who also need those services. And, as we saw in the case of Saragosa, cultural barriers such as language differences may hamper effective response measures, with tragic results (Phillips, Garza, and Neal, 1994). By identifying such problems, research on social diversity and inequality can promote the development of more effective plans for emergency response and recovery.

DISASTER AGENT CHARACTERISTICS

Disasters vary along a number of dimensions including frequency, familiarity, duration, severity, scope of impact, destructive potential, and the length of the warning period they permit. These attributes have important consequences for planning and response. Consider, for example, the implications of the ability to forecast disaster impact. A disaster with a longer warning period makes it possible to issue warnings to the public and to increase response capability—for example, by notifying potential respond-ers of the threat and moving emergency vehicles and equipment into pre-designated staging areas and safe locations. A significant warning period allows threatened communities to engage in efforts to mitigate damage, such as boarding up windows and tying down objects. Other things being equal, then, we would expect response activities to be more effective and losses to be reduced in disaster situations in which warning is possible.

With respect to scope of impact, perhaps the simplest distinction that can be made is between communitywide or even regional disasters and those that are relatively localized and site-specific. In general, re-search suggests that the smaller the scope of impact the more amendable the situation is to effective management, for several reasons. First, disas-ters that strike a limited area typically do not destroy or disrupt as many community resources as those with a larger area of impact. When lim-ited-scope disasters leave the infrastructure intact, communication, trans-portation, and other response-related tasks are less difficult. When a di-saster has a relatively focused impact, there are typically large numbers of unaffected residents in the community who are able to provide assis-tance, and both damage assessment and control over ingress and egress at the disaster site are less difficult. Large-scale, communitywide disas-ters typically create more significant problems for affected communities. Infrastructure damage may be widespread, disaster response resources may themselves experience damage and disruption, and damage assess-ment and the management of sites where severe impacts have occurred are more problematic. Additionally, larger events almost invariably in-volve the mobilization of emergency personnel from other local jurisdic-tions, the activation of mutual aid agreements, and the participation of state, regional, and federal agencies, which expands the need for inter-organizational and intergovernmental coordination. Other things being equal, then, large-scale, multi-jurisdictional events are probably more difficult to manage effectively than more localized ones.

Hazards also differ in the extent to which they are familiar to com-munity residents and emergency responders. Familiarity is generally a function of the degree of prior experience a community has had with a particular disaster agent. As noted above, experience can lead to both desirable and undesirable outcomes. On the one hand, experience may make disasters or particular disaster agents more salient to community residents and local officials, stimulating stepped-up preparedness and response efforts. On the other, it may engender complacency or fatalism. Additionally, because communities have a tendency to plan for the disasters that are most frequent (and thus most familiar), in the process they may neglect low-probability, high-consequence events.

Natural Versus Technological Disasters

Since the time of the last assessment, a major debate has developed on the issue of whether natural and technological disaster agents differ in terms of the preparedness and response behaviors they stimulate. Several perspectives on this issue appear in the literature. Influenced in part by the aftermath of catastrophic events like the Three Mile Island and Chernobyl nuclear accidents, the Bhopal disaster, and the Exxon oil spill, one body of research suggests that disasters involving technological agents constitute a distinct genre—that is, that the social and behavioral patterns that occur in emergencies and disasters involving technological agents differ from those occurring in natural disaster situations. Those making this distinction are usually not referring to the entire range of technologically-induced accidents and disasters—for example, transportation-related incidents like train wrecks and airplane crashes are typically not included—but rather to those that involve unfamiliar, exotic, and dreaded hazards such as nuclear power, nuclear waste, and dangerous chemical substances.

Technological disasters are considered distinctive in several ways. Some studies suggest that technological disaster agents produce responses in the public that differ from what commonly occurs in natural disaster situations. For example, it is normally difficult to bring about compliance with disaster warnings; people show a general tendency to normalize, to discount threat messages, and to seek confirmation before considering action. In contrast, some researchers argue that almost the opposite occurs in technological emergencies, particularly in situations involving nuclear hazards. Following the Three Mile Island nuclear accident, for example, large numbers of people left the area even though they had not been told formally to evacuate (Flynn, 1982; Stallings, 1984).

The two types of agents are also thought to vary in their short- and longer-term impacts. For example, while natural disasters are widely believed to produce few discernable serious or longstanding mental health problems in victim populations, some studies report heightened levels of psychological distress in the wake of nuclear accidents and other technological emergencies (Dohrenwend et al., 1981; Baum, Fleming, and Davidson, 1983; Smith et al., 1986; Houts, Cleary, and Hu, 1988; Picou et al., 1992; for a more extensive review see Freudenburg and Jones, 1991).

These negative psychosocial impacts are attributed both to the distinctive properties of technological agents and to the community problems they produce. For example, while people understand the experience of going through a tornado and what the effects are likely to be, they may be worried and anxious about the longer-term impacts of toxic chemicals, both to themselves and to the environment. The uncertainty and ambiguity that accompany exposure to technological hazards are thus thought to heighten stress.

As we suggested earlier in our discussion of altruism during the disaster response phase, natural disasters are generally characterized as "consensus" crises (Quarantelli and Dynes, 1976) that are accompanied by heightened community cohesiveness and morale and by the emergence of a "therapeutic community" (Barton, 1969) that helps victims cope with loss and facilitates reintegration and recovery. In contrast, technological hazards are thought to result in heightened levels of community conflict—conflict that is reflected in the subsequent emotional and adjustment problems residents report.

Couch and Kroll-Smith (1985), originally proponents of the idea that natural and technological disasters differ in their effects, developed the concept of the chronic technical disaster to characterize slow-onset technologically-induced threats like Love Canal and the Centralia, Pennsylvania, mine fire that have unclear longer-term consequences and that create conflict over their causes, the nature of the threat they pose, and how to undertake ameliorative action. Their position was that such hazards differ from natural hazards in that they generate hostility and mistrust directed toward the agents that produced the hazard and the agencies responsible for remediation, as well as among community groups. Such conflict is likely to have a lasting corrosive effect, leaving individuals, neighborhoods, and communities worse off over time (see also Cuthbertson and Nigg, 1987). Pijawka, Cuthbertson, and Olson (1987–1988) discuss a number of reasons why technological threats are particularly likely to produce negative and divisive—as opposed to positive and unifying—effects on communities. Those reasons include anxiety about the long-term effects of exposure, the element of human blame, the fact that entire communities or segments of communities may be affected, the tendency of affected people and neighborhoods to be stigmatized, and the tendency for political conflict to develop over how to deal with technological hazards.

Some of the claims made to support the natural/technological distinction are arguable. For example, Kasperson and Pijawka (1985: 8) have distinguished natural and technological hazards in the following ways:

> The hazards of technology pose different managerial problems than those arising from nature. Natural hazards are familiar and substantial, accumulated trial-and-error responses exist to guide management; technological hazards are often unfamiliar and lack precedents in efforts at control . . . Natural hazards tend to provide only limited potential for preventing events, and, thus, management tends to occur "late" in the hazard chain . . . Members of the public tend to see natural hazards as acts of God whose effects can only be mitigated; technological hazards . . . are assumed to be amenable to "fixes" of various kinds, and amenable to substantial reduction.

This statement is accurate for some natural and technological disaster agents, but not for others. Some types of natural hazards—catastrophic earthquakes and major volcanic events, for example—are quite unfamiliar at the community and even societal level. Some technologically-induced incidents, such as chemical releases in communities with high concentrations of processing facilities, may be almost routine. Some natural disasters can indeed be prevented, or their effects greatly reduced, through such measures as sound building practices and effective land-use policies and regulations.

Earlier work by Couch and Kroll-Smith argued (1985: 566) that "[t]he chronic technical disaster develops slowly and persists for a relatively long time" and that "while the effects of natural disasters are often influenced by human factors . . . chronic technical disasters are *caused* by human-technological intervention in the environment, and further technical human intervention is required to *contain* or abate the disaster agent itself." However, some natural disaster agents, such as droughts, also develop slowly and last a long time. Volcanoes are accompanied by long warning periods, they keep erupting over a relatively long period of time, and their effects can also be quite long lasting. Aftershocks following major earthquakes can last for months and even years. Natural disasters, in other words, also have a "chronic" side.

The distinction between human intervention "influencing" and "causing" disasters is also a matter of degree. Following Hurricane Andrew in 1992, for example, analyses of what caused the losses in that multi-billion-dollar event placed almost equal emphasis on the intensity of the storm and on building practices, lax building code enforcement, and Southern Florida's development and coastal zone management policies. Technical interventions such as constructing dykes and levees and retrofitting buildings are also employed to contain or abate the effects of natural disasters. Indeed, because such interventions can be costly, and also because they often involve mandates and regulations, they can spark just as much conflict and controversy as the measures that are employed to deal with technological threats. Again, looked at more closely, the differences between natural and technical hazards are not so large.

In contrast with the commonly-used natural/technological distinction, some analysts suggests there are other underlying dimensions that do a better job of explaining differential responses to hazards. Thus, a second position in the natural vs. technological debate is one that employs a common system for classifying all types of disasters. Some scholars argue that the key to understanding how people and organizations respond lies in the more general characteristics of disaster agents, such as familiarity, speed of onset, length of the warning period, and scope of impact. Perry and Mushkatel (1984), for example, studied the response of both white and minority community residents to warnings involving a flood and a hazardous materials incident. They found that the same overall model of evacuation compliance explained behavior in both emergency situations. Differences were observed (Perry and Mushkatel, 1984: 217) both among the different ethnic groups studied and between the two events, but those differences were due not to the natural/technological distinction but rather to event characteristics on which both natural and technological agents vary:

> [c]itizen evacuation performance in Railtown [the hazardous materials incident] was similar to that in other threats—including floods and volcanic eruptions . . . characterized by low levels of disaster event forewarning, comparatively short time periods for citizen warning compliance, and a non-complex adaptive strategy . . . closely overseen by officials.

Thus, one alternative to arguing for a strict natural/technological distinction is to focus on more abstract properties of disasters that create differential responses among affected publics, organizations, and communities. Natural and technological agents that resemble one another along particular dimensions—for example, that occur without warning, have a large scope of impact, are unfamiliar, or are of long duration—will produce similar kinds of responses and challenges.

Similarly, Kroll-Smith and Couch, who once took the position that technological disasters are distinctive in their effects, later changed their perspective to one that emphasizes the interpretive processes through which individuals and groups assess hazards and disasters. According to this revised view—which they term the "ecological-symbolic perspective"—understanding how different social groups assign meaning to hazard-related experiences and what those meanings are is the key to understanding the impacts those experiences have. In their view (Kroll-Smith and Couch, 1991: 362):

> the issue that separates types of aversive agents is not simply whether they are natural or technological. More importantly, what is the differential impact of the agent on the built, modified, or biophysical environments, and how are these impacts experienced?

Despite such efforts at clarification, a division persists in the disaster literature between the "generic" and "event quality" perspectives (Kroll-Smith and Couch, 1991). Exemplifying the former position are Quarantelli and Dynes, who have long argued that disasters should only be described and understood in social rather than physical terms (see, for example, Quarantelli, 1992). Exemplifying the latter are researchers like Freudenburg and Jones, who find strong support in their own and others' work for the notion that "[t]echnological disasters have been associated with an increasingly broad range of negative—and strikingly long-lived—social and psychological impacts" (1991: 1154). This position is echoed by Erikson (1994), who describes technological emergencies like Three Mile Island as a "new species of trouble" that traumatize individuals, undermine community solidarity, and destroy the public's trust in its institutions.

A third perspective that has recently begun to appear in the literature argues that features of disasters like those discussed here—for example, whether an event is natural or technological, whether a particular party is to blame, and whether disaster impacts can be prevented or mitigated—can usefully be viewed as social constructions. According to this view, taking a position that particular types of disaster agents are accompanied by particular types of social behavior—fearful responses, conflict, or blaming—ignores the fact that people respond to phenomena in terms of the meanings they assign to them. For example, Stallings's recent analysis of the earthquake problem (1995) argues that the earthquake threat is socially constructed, the product of promotion and claims-making by a group he terms the "earthquake establishment." Stallings is not, of course, taking the position that earthquakes are not real and damaging. Rather, his study documents the ways in which organized social actors frame the earthquake problem as a putative threat and describes the social processes involved in the formulation and adoption of recommended "solutions" to the earthquake problem.

In social-constructionist work that is directly relevant to the debate over the natural/technological distinction, Blocker and her colleagues have shown how, rather than viewing a flash flood that struck their city as a naturally-occurring event, many Tulsa residents saw it as "resulting from a lack of control over technological systems that could, and should, have protected them from harm" (Blocker, Rochford, and Sherkat, 1991: 368), and they blamed the Army Corps of Engineers and local government for faulty management of the flood-plain and flood-control system. A local protest movement developed to oppose flood-plain remapping, challenge development policies, and pressure for better mitigation. In this case, the flood, which under other circumstances might have been considered a natural event, was collectively defined as human-caused, and protests not unlike those that accompany some technological hazards occurred.

Blocker and Sherkat (1992) have also argued that interpretations of disasters are increasingly shifting to define them in general as originating in technology—that is, in the failure of humanly-devised systems of control. These interpretations in turn shape judgments about responsibility for reducing disaster losses:

> Once-natural risks may be joining the growing list of technological risks which are considered to be avoidable results of industries' or governmental agencies' sins of omission or commission. Further, actual increases in technological sophistication suggest to us that virtually all calamities, whether they result from natural processes or human machinations, could conceivably be avoided (1992: 164).

In short, rather than being defined as "acts of God," the cultural trend is toward seeing disasters of all kinds as "acts of man."

Aronoff and Gunter's (1992) constructivist study of one community's response to toxic contamination of a community challenges the idea that chronic technological threats invariably damage morale and cause community conflict, pointing instead to the ways in which socially-generated meanings guide actions taken with respect to hazards. Rather than reacting negatively when reports of extensive PBB contamination surfaced, as the literature on technological threats would predict, community residents redefined the problem as one requiring a unified effort to obtain cleanup resources while keeping the community's economic base viable. Even though toxins were involved, the social definitions they developed enabled them to approach the PBB problem much as they would a natural disaster event.

This case demonstrates that how threats and events are socially constructed has a greater influence than their presumed inherent characteristics on how people subsequently respond. By showing how views of disaster events, their causes, and their consequences are shaped by social, cultural, and institutional practices, a social-constructionist approach effectively undermines the natural/technological distinction. Equally important, this perspective leads logically to a consideration of how and by whom such constructions are developed (Clarke and Short, 1993).

Another slightly different point of view is that the natural/technological distinction is an artificial one because human agency is the key factor in all disaster events. Blaikie et al. (1994: 6) argue, for example, that "[t]he 'natural' and the 'human' are so inextricably bound together in almost all disaster situations, especially when viewed in an enlarged time and space framework, that disasters cannot be understood to be 'natural' in any straightforward way" (see also Wijkman and Timberlake, 1988). Similarly, Tinker's article, entitled "Are Natural Disasters Natural?" (1984) distinguishes disasters which are the products of human activity from the events in nature that trigger them. This approach to the disaster problem leads to the following analysis (1984: 14) of the causes of devastating floods that occurred in 1981 along China's Yangtze River:

> Deforestation in the Yangtze catchment started early in the last century, when trees were cut for imperial palaces. The process accelerated during China's disastrous 'Great Leap Forward' toward rapid decentralized industrialization, which included a campaign to establish backyard iron furnaces that needed wood for fuel. In the 1960s more forests were felled for the 'grain first' campaign.

In other words, rainfall may have been the trigger for the flooding, but the flood *disaster*, an outcome directly traceable to deforestation, was socially generated.

In summary, the question of whether particular kinds of technological disaster agents produce responses and outcomes that are different from those observed in natural disaster situations remains a subject of dispute. The field remains divided, with some researchers arguing for a generic approach to disasters and their impacts, others holding an almost essentialist view that technological agents have especially pernicious effects, and still others taking the position that the origins and characteristics of disaster agents are largely socially constructed. This paradigm clash has the potential for both stimulating further research on disasters and their consequences and for improving our ability to theorize about their origins.

CONCLUDING OBSERVATIONS

Since the last assessment of hazards research, major strides have been made toward better understanding and explaining individual, group, and organizational preparedness and response activities. At the same time, scholarship in the field has shifted from seeing the behaviors of affected populations as more or less homogeneous or undifferentiated to recognizing the heterogeneity of those behaviors. The public is increasingly being characterized as internally differentiated, consisting of subpopulations whose experience with hazards and disasters varies as a function of gender, ethnicity, social class, disaster-related knowledge and experience, and other socioeconomic and sociocultural factors. The interplay of these factors could become a major focus for research, supplanting classic studies that concentrated more on what residents of vulnerable or disaster-stricken communities had in common than on how they differed. The image of the consensus-building and status-leveling forces that come to the fore in disaster situations has been replaced by an alternative view. That newly-emerging framework places equal emphasis on the fact that broad agreement on the need to preserve life and safety in the aftermath of disaster can coexist with inequality and that, while bringing people together, disasters can also further marginalize already marginal populations.

It has long been argued that there is considerable continuity between pre-disaster conditions and trends and what occurs when a disaster strikes. Disasters disrupt the social order but they do not obliterate it, and while disasters may accelerate pre-disaster trends they rarely reverse them. One clear implication of this continuity principle is that everyday patterns of social inequity—such as unequal access to housing, information, services, and political power—carry over into post-disaster settings and are reflected in victims' experiences. Hazards and disasters do not ameliorate the problems that stem from inequality and poverty; instead, they can exacerbate them. This idea applies to structurally-based differences among both pre-disaster activities and post-disaster behaviors and outcomes. Similarly, as the research discussed here has shown, it applies equally across other social units, in that less-well-off communities and organizations likely experience the same sorts of disadvantages as less-well-off households.

CHAPTER SIX

The Wider Context: Societal Factors Influencing Emergency Management Policy and Practice

T HE APPROACHES USED BY different societies to manage hazards and disasters are in large measure a reflection of the distinctive characteristics of those societies. Preparedness and response activities take place within particular governmental systems and are shaped by larger cultural, economic, and political forces. Taking these broader societal factors into consideration can shed light on the manner in which hazard management activities are organized and the reasons why particular hazard adjustments are preferred over others. It can also help explain why some approaches to loss reduction succeed in particular societal settings while others fail and still others are never considered at all. Additionally, situating hazard and disaster management policies in their societal contexts can lead to a better understanding of the extent to which both research findings and policies and practices can be generalized from one society to another.

In the previous chapter we reviewed the literature in order to show the ways in which a variety of social, economic, and agent-related factors influence preparedness and response activities among different social units. This chapter takes a more macro-social view, focusing on how the characteristics of the U.S. governmental system and of American society in general have shaped the hazard and disaster management strategies of both governmental units and the general public. Among the topics discussed are governmental organization and its implications for emergency management; the impact of mandates and incentives; how cultural ideas and beliefs influence the assumptions we make about how to manage hazards; the role economic forces play in structuring choices in the hazards area; and the impact of major trends such as technological change and the professionalization of emergency management.

THE INTERGOVERNMENTAL AND POLICY CONTEXT

The organization, effectiveness, and in particular the tremendous diversity of preparedness and response efforts in the U.S. are in large measure a consequence of the structure of hazard management policy, which is in turn embedded in the manner in which the broader intergovernmental system operates. Studies of American hazard management policy have highlighted two factors that complicate the conduct of hazard reduction activities. First, responsibility for different aspects of the disaster problem is diffused among many agencies at different governmental levels. Second, authority relationships among those agencies are weak, which impedes implementation and lessens accountability. As Sylves has pointed out (1991: 416), local governmental politics in the U.S. are bewilderingly complex, with "3,000 counties, 16,700 townships, and 29,000 special districts, each with their own policy making structure." Writing on the topic of disaster preparedness, Waugh noted that (1988: 118–119):

> [t]he federal system itself acts to inhibit coherent and comprehensive disaster preparedness efforts. Vertical fragmentation due to the division of powers between the federal and state governments and the limited powers given to local governments by states make decisionmaking and program coordination awkward at best and ineffective at worst. Horizontal fragmentation due to the jurisdictional prerogatives of a multitude of agencies adds to the difficulties.

The emergency management system in the U.S. has been influenced to a very significant degree by the intergovernmental system's jurisdictional complexity, overlapping and often inconsistent authorities, and the high degree of power held by local jurisdictions (Sylves, 1991). May and Williams (1986: 180) have documented how this system of "shared governance" disperses disaster-related responsibilities among various agencies and levels of government, generating fundamental tensions:

> On the one hand, the federal interest is in increasing both subnational commitment to federal goals and the capacity of subnational governments to carry out regulatory or programmatic activities in support of those goals. On the other hand, the predominant subnational interest is in having access to federal resources and expertise, *and* in having sufficient discretion to reshape the federal goals or specific means in light of special needs or political concerns.

Hazard management policies have also been altered in recent years by major changes in the broader intergovernmental system. These changes have consolidated some functions at the federal level, but also have tended to give more autonomy to states and local governments, made the policy environment more uncertain, and complicated implementation efforts.

Analyses by scholars studying the history of emergency management in the U.S. (Kreps, 1990; Drabek, 1991b) make the point that hazard policy has been fragmented since its inception. Prior to 1950, Congress had passed over one hundred separate pieces of legislation to provide different forms of disaster relief, but the passage of the Federal Disaster Act that same year represented the first attempt to establish an ongoing system for disaster relief. Disaster policy changed and the federal role expanded following subsequent disasters, including the 1964 Alaska earthquake, hurricanes Betsy, Camille, and Agnes (1965, 1969, and 1972, respectively), and the 1971 San Fernando Earthquake (Drabek, 1991b; National Academy of Public Administration, 1993). However, legislation directed toward problems resulting from these and other disaster events contributed to further diffusion of responsibility.

This situation was compounded by a persistent institutional ambivalence regarding the federal role (and, by implication, the roles of other governmental levels) in emergency management and what its goals should be. Specifically, should such efforts be directed primarily toward wartime readiness (for example, preparation for a nuclear war) toward civil disasters, or both? Initially, federal efforts tilted very much in the direction of war-related preparedness. In 1950, the same year the first general disaster law was passed, the Federal Civil Defense Administration (FCDA) was established within the Executive Office of the President (EOP); shortly thereafter, the Federal Civil Defense Act of 1950 gave the FCDA independent agency status. At about the same time, the Office of Defense Mobilization was created, again within the EOP. In 1953, the

NSRB and ODM were combined into the Office of Defense Mobilization. In addition to the ODM another office, the Federal Civil Defense Administration (FCDA), played an important role in war-related crisis planning.

As shown in Figure 6.1, which is adapted from Drabek's (1991b) analysis of how federal emergency management policy has developed over the past five decades, both wartime planning and federal disaster policy underwent continual reorganization throughout the 1950s and 1960s. The Defense Department's Office of Civil Defense (OCD, later to

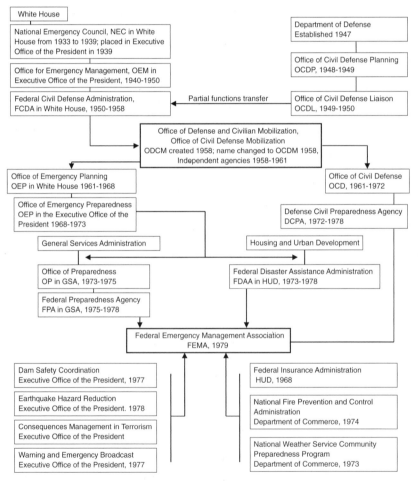

Figure 6.1 Evaluation of Federal Disaster Management Functions (From Drabek, 1991b. Used with permission)

become the Defense Civil Preparedness Agency) was established. Among its roles was to assist states and local governments in preparedness for nuclear war. The Office of Emergency Preparedness (OEP) was also given a major role in defense planning. Disaster-related responsibilities were divided among the Federal Preparedness Agency (FPA) in the General Services Administration, the Federal Disaster Assistance Administration (FDAA) within HUD, and other agencies. A federal reorganization in 1973 resulted in a division of federal emergency management responsibility among three agencies: FPA, FDAA, and DCPA.

The Federal Emergency Management Agency (FEMA) was formed in 1979 in an effort to overcome the fragmentation that had developed in hazard management policy and programs. FEMA brought together five federal agencies with key crisis-management roles: the DCPA from the Department of Defense; the FPA from the General Services Administration; the FDAA and the Federal Insurance Administration from Housing and Urban Development; and the National Fire Protection and Control Agency from the Department of Commerce. However, the reorganization also left a number of specialized programs in other agencies. These included the Army Corps of Engineers, the Small Business Administration Disaster Loan Program, radiation protection programs in the Environmental Protection Agency and the Nuclear Regulatory Commission, the hazardous materials programs in the Department of Transportation, and the Department of Energy's nuclear hazards program (Kreps, 1990).

The creation of FEMA achieved a significant degree of consolidation in the management of federal disaster programs. It also began a trend toward considering disaster-related issues on an equal footing with those associated with wartime emergencies. However, reorganization was not without its own problems. Crisis relocation planning, a defense-related program, was resisted strongly by many local jurisdictions (May and Williams, 1986). Equally important, FEMA reorganization could not address the lack of vertical integration that results from our system of shared governance, among the main consequences of which is that "no single level of government can capture control of the entire policy process" (Lindell and Perry, 1992).

The U.S. policy system assigns local government the primary responsibility for emergency management, but this arrangement has a number of drawbacks. Community emergency management networks vary considerably in their organization, resources, and overall effectiveness. Local governments have less revenue-generating capacity than do other governmental levels, and, typically, disaster-related issues must compete with other concerns that are much higher on a local community's agenda. Too often, emergency management is assigned a low priority at the local government level.

Very little research exists on the ways in which governmental structure and policies influence preparedness and response activities, and most of that work has been done only in the U.S. The few non-U.S. and cross-national studies that exist do little more than suggest what those impacts might be. Some work explores the impact centralization of government functions, as opposed to decentralization, has on the handling of disaster situations. McLuckie's (1977) research compared disaster responses in Japan and Italy, which at the time were more politically-centralized nations, with those of the U.S. on the assumption that centralization of political authority would have an impact on how response activities were carried out. After taking into account other factors such as societal differentiation and the level of technological development, McLuckie found that governmental centralization did affect the performance of disaster-related functions, but that those effects varied depending on disaster phase and the tasks performed. For example, centralization was more common during the pre-disaster planning phase than during the emergency response period and less common for tasks such as evacuation and victim care.

Dynes, Quarantelli, and Wenger's (1990) study of the governmental response to the 1985 Mexico City earthquake also suggests that the way political authority is structured during non-disaster times affects the manner in which disasters will be handled. Authority was not centralized prior to the Mexico City disaster, and the decentralized response structure that emerged following the earthquake was a continuation of that pattern.

Some studies suggest that approaches to disaster preparedness and response (as well as those concerned with mitigation and recovery) are also influenced by political ideologies and patterns of institutional dominance within societies. For example, in the developing countries of Latin America and Africa disaster mortality rates are lower in more egalitarian societies than in countries ruled by authoritarian regimes supportive of economic elites. Correspondingly, these "corporatist" regimes emphasize protecting infrastructure and economic resources over protecting people; as a result, their disaster-related property losses are lower (Seitz and Davis, 1984).

More research is needed to better understand the ways in which differences in governmental and state systems affect hazard and disaster management. Only a handful of studies have dealt with the role of state structure in the management of risks, or with cross-national differences in hazard-management policies. For example, Brickman, Jasanoff, and Ilgen (1985) studied the politics of controlling chemical hazards in the U.S., Great Britain, France, and Germany, and Jasper (1990) analyzed nuclear energy policies in the U.S., Sweden, and France, but systematic comparative research on natural hazard management is absent from the literature.

The Influence of Mandates, Incentives, and Program Guidelines

Higher levels of government use a variety of mechanisms to influence activities at lower levels. Those mechanisms include incentives, or rewards that are given for voluntary compliance; mandates, or rules for which violation is punished; and guidelines that accompany funds and other types of assistance. Although the sizeable literature on implementation shows clearly that merely passing laws and issuing directives does not ensure that desired changes will actually take place, there is evidence that mandates and other types of legal and regulatory requirements can have a positive impact on disaster preparedness and response, especially if they are applied with consistency and accompanied by evidence of serious commitment (May and Williams, 1986). Indeed, while the kinds of disaster plans many local communities have developed and the approaches they have taken to preparedness may leave much to be desired, it is unlikely that formal disaster plans would have become almost universal at the local level if they had not been required. Similarly, we may be critical of the evacuation plans that have been developed for nuclear plant emergency planning zones and of the nuclear emergency planning process in general, but it is also clear that changes in the regulatory environment after Three Mile Island did spur nuclear facilities and the communities surrounding them to expand their planning efforts (Sylves, 1984). The fact that holding regular disaster drills is a requirement for continued certification by the Joint Commission for Accreditation of Health Care Facilities is no doubt one important reason that hospitals have those exercises. Without SARA Title III, it is unlikely that local communities with hazardous materials facilities would have developed multi-organizational networks to deal with chemical emergency planning and response. Compliance with the law is still far from complete, but its passage did speed the emergence of new preparedness networks (Lindell, 1994a). Research conducted prior to the enactment of that legislation shows that such networks did not exist, even in high-risk areas (Quarantelli, 1984).

Research on hazard mitigation also suggests the importance of policies that require the adoption of hazard-reduction measures, rather than allowing them to be undertaken voluntarily. Examples include the adoption of building codes and hazardous building retrofit ordinances, the requirement that local communities in California adopt seismic safety elements for their general plans, and the federal flood insurance program. Of course, none of these programs has been completely effective and, as with programs designed to deal with other kinds of social problems, implementation is always problematic. However, without them it is likely that even less hazard-reduction activity would have occurred.

Despite the fact that mandated programs have caused improvements, it is equally clear that not all such programs achieve their objectives, because merely passing laws and issuing directives does not ensure that desired changes will actually take place. Indeed, there is evidence that poorly-designed programs actually can slow down hazard reduction efforts because they generate resistance rather than compliance. This is particularly clear in May and Williams's (1986) analysis of the "degenerated" collaborative activity that accompanied FEMA's ill-fated crisis-relocation program. Moreover, mandates are typically resented by the governmental units at which they are directed, and unfunded mandates are especially unpopular. Unless funds exist to actually carry out programs, mandates alone cannot be expected to have an impact.

While most national-level programs have focused on mitigation rather than preparedness and response, the federal government expended some effort on encouraging subnational levels of government to improve their disaster management capabilities. While not mandates in the strict sense, these approaches have attempted to provide "guidance" and to specify performance standards and goals. Working for the National Governors' Association (NGA) under the sponsorship of the Defense Civil Preparedness Agency (FEMA's precursor) Whittaker developed a concept she termed "comprehensive emergency management," which was intended to encompass all phases of all types of disaster events using a comprehensive planning approach. This comprehensive planning concept was described in an NGA publication entitled *Comprehensive Emergency Management: A Governor's Guide* (National Governors' Association, 1979). Not long after it was formed in 1979, FEMA adopted this planning approach, renaming it the Integrated Emergency Management System (IEMS), with the goal of moving local jurisdictions further toward comprehensive hazards analysis and heightened disaster management capacity. IEMS, which was FEMA's first attempt to encourage an all-hazards approach that would encompass all four phases of the hazard cycle—mitigation, preparedness, response, and recovery—envisioned a multi-stage hazard management process involving analyses of community vulnerability; assessment of local emergency management resources and capabilities; development of comprehensive emergency plans; maintenance of response capability through training and other activities; and improved response and recovery management (McLoughlin, 1985; Sylves, 1991). Detailed documents were prepared providing directives on what local governments should do in order to develop the integrated systems FEMA envisioned (see, for example, Federal Emergency Management Agency, 1983). However, implementation of the IEMS concept was incomplete and uneven for various reasons, including:

> [d]ifficulties inherent in U.S. intergovernmental relations, the weak institutional status of emergency management agencies (especially at the national level), fragmentation of disaster/crisis responsibilities at each level of government, weak political constituencies advocating improved emergency management, [and] severely constrained national budgetary authority (Sylves, 1991: 423).

Other more recent federal strategies for enhancing emergency management capability and increasing accountability include performance partnership agreements (PPAs), Comprehensive Collaborative Agreements (CCAs), and reporting systems such as the Computerized Activities Results List (CARL) and the Capability Hazard Identification Program (CHIP). These intergovernmental arrangements sought to require states to undertake specific tasks, reach benchmarks, and report on their activities as a condition for receiving federal funding. Little systematic research has been done on how these programs operate in practice or on the extent to which their goals are being met. The available evidence suggests that these programs certainly are preferable to having no monitoring or accountability at all. Yet existing systems need improvement, both in stimulating other governmental levels to work harder on hazard-related problems and in measuring the effectiveness of hazard reduction programs. For example, in their assessment of activities undertaken by FEMA with participating states under the National Earthquake Hazards Reduction Program, Gillespie et al. (1995) found that criteria for state performance tend to be vague; that record keeping and reporting are often incomplete and inconsistent, making it difficult to assess program activities, both at given points in time and over the life of the program;

and that the data that are provided are insufficient to evaluate program impact.

FEMA's own Inspector General's office reached similar conclusions when it conducted a study of the CCA process (Federal Emergency Management Agency, 1994). That report found that the different levels of risk states face were not taken into consideration in the granting of funds and that the two main reporting systems that were designed to monitor state activities, CARL and CHIP, were not set up and used in ways that made it possible to assess what states are actually doing to improve their ability to handle disasters. The report was particularly critical of burdensome reporting requirements, such as lengthy "crosswalks" and checklists, that are very labor-intensive but that nevertheless do not provide needed information on the operational capacity and performance of emergency management organizations. Missing from these efforts at ensuring accountability is an understanding of what is really going on at state and local levels in the areas of preparedness and response. The report (1994: 13) concluded bluntly that:

> FEMA currently does not assess the emergency management capability of the states. The two management information systems that FEMA could use [CARL and CHIP] are inadequate. Furthermore, the way exercises are managed and reported does not tell how states will respond to disasters.

On the basis of such reports, it appears that the federal government has not yet developed a strategy for using mandates and regulations to influence the way emergency preparedness and response activities are carried out at subnational levels. Although legally states are required to follow certain procedures, such as developing emergency response plans and providing training for emergency management personnel, existing accountability mechanisms are only of limited use in improving preparedness and response because they do not address issues of substance.

The Gillespie and Inspector General reports suggest strategies the federal government could adopt that might be more effective in bringing about desired improvements. These include taking risk into account in allocating funds for preparedness and response activities, negotiating standards for actual performance with states, and following up more thoroughly to see whether agreed-upon actions were taken. Moreover, performance audits should focus on outcomes—that is, actual changes that have taken place and impacts programs have had. Assessment procedures should be made consistent across jurisdictions and over time, so that meaningful comparisons can be made and progress toward goals can be tracked. Finally and most significantly, the federal government should evaluate both disaster exercises and actual disaster response activities. In short, the government has more tools at its disposal than it is currently using, and it needs to shift its emphasis from the evaluation of paper plans to a broader assessment of emergency management capability that addresses issues of staffing, training, equipment, facilities, and intergovernmental coordination.

Other Efforts to Influence Emergency Management Practice: The Incident Command System and the Federal Response Plan

Historically, approaches to the management of emergencies and disasters have been both community-specific and organizationally idiosyncratic. Rarely have two systems of crisis management been even similar. By the early 1980s, however, the fire services in particular became concerned that fire departments needed a common command system to enhance their effectiveness in responding to larger incidents. This problem was strongly felt in Southern California, where wildfires routinely required the coordinated response of many fire departments from various jurisdictions. Departments began working together to plan large-scale responses, and with funding from FEMA a program called FIRESCOPE (Firefighting Resources of Southern California Organized for Potential Emergencies) was developed (for a description of this program see Federal Emergency Management Agency, 1987). One key element in the FIRESCOPE planning and operations model was a component that focused on the management of firefighting operations, known as the Incident Command System (ICS). ICS characterized the management of crisis incidents as involving four components: operations, logistics, planning, and finance. It also aimed at reducing ambiguity about lines of authority in emergencies by assigning responsibility for incident management to the agency representative who is first on the scene when an emergency develops.

For several years, FIRESCOPE and ICS were used primarily for very large, multi-jurisdictional fire incidents rather than for routine fire emergencies. However, the ICS model was later revised by Brunacini (1985) and made applicable to smaller fire emergencies in addition to larger ones. Brunacini also changed the incident command function to include specialized advisors, expanded the operations function to include routine fire department response demands, such as the deployment of hazardous materials teams, and incorporated explicit connections to emergency operations centers and police agencies. Figure 6.2 shows a typical fire department incident management structure. ICS and its variant, the Incident Management System (IMS), are now routinely used in the American, Canadian, British, and Australian fire services, and both university-based fire services programs and the National Fire Protection Association provide instruction in the use of these management models.

ICS has increasingly come to be seen as appropriate for the management of emergencies other than fires, including both natural and technological disasters. ICS and its four-function crisis-management framework have now been widely diffused among emergency management agencies and other crisis-relevant organizations both at local and supra-local levels. In some cases, use of the system has become mandatory. For example, the State of California recently passed legislation requiring all local jurisdictions to adopt a variant of ICS it calls the Standardized Emergency Management System (SEMS). This measure was seen as a means of eliminating the confusion that had existed over how to organize and manage disaster operations, while at the same time enabling many jurisdictions and agencies to integrate their activities more smoothly during disasters.

During the mid-1980s, the federal government also began a comprehensive planning effort in order to allocate tasks and roles among the

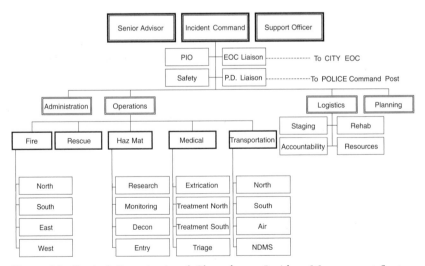

Figure 6.2 Typical Organizational Chart for an Incident Management System (Adapted from Brunacini, 1985)

various federal agencies that had responsibilities and legal authorities in the area of disaster response. An interagency planning initiative was thought to be particularly important for the management of catastrophic events, such as the very large earthquakes that could occur in the central U.S., greater Los Angeles, and the San Francisco Bay region. This effort ultimately resulted in the development of the Federal Response Plan (FRP), which aims at mobilizing the resources of 26 different federal agencies and the Red Cross in federally-declared disasters. The FRP is organized around the performance of 12 "emergency support functions" (ESFs), which include key response tasks such as providing shelter, food, and health and medical services to victims, as well as other activities such as communications, transportation, and information-gathering and planning, that are central to the handling of response operations. For each ESF, one federal agency is assigned primary responsibility, with others designated as providing support to that agency.

The FRP was first implemented in its entirety following Hurricane Andrew in 1992. While the intergovernmental response to that disaster was widely judged to be inadequate, the panel that was subsequently set up to evaluate the governmental disaster response system called the FRP "an important beginning" (National Academy of Public Administration, 1993: 30) in what should be an ongoing process of improving federal emergency management policy, noting "that it [the FRP] exists at all is a credit to FEMA . . . " (National Academy of Public Administration, 1993: 29).

Both ICS and the FRP are designed to clarify key response-related tasks and overcome the confusion that invariably develops when multiple agencies and jurisdictions mobilize during major disasters. Both endeavor to impose a consistent organizational structure, task breakdown, and terminology on a set of activities that were known to show great variability across responding organizations, communities, states, and disaster events. And both are based on the assumption that pre-planning and well-understood lines of communication and responsibility yield better disaster management. However, there has been no systematic empirical research on the effectiveness of either ICS or the FRP as organizing mechanisms for disaster response. It is also important to recognize that, while there have been a number of very serious disasters in the U.S. during which the FRP was activated and judged to be effective, and while occasional exercises are conducted based on even larger disaster scenarios, the nation has yet to experience the type of catastrophic event for which the FRP was originally developed. Thus, while it seems intuitively

correct to expect that federal planning efforts will improve disaster response in truly severe disasters, it still remains to be seen.

CULTURAL INFLUENCES ON HAZARD- AND DISASTER-
RELATED BEHAVIOR

The concept of culture encompasses the values, beliefs, assumptions
about the world, and distinctive behavioral practices that groups and
societies share. Important cultural elements are generally reflected both
formally, in laws and regulations, and informally, in customs and the
behavioral expectations held by members of a society or group. Cultural
expectations and practices inform hazard-related behaviors and practices just as they do other aspects of social life. Cultures differ, for example, in ideas about risk-taking, individual versus collective responsibility for loss reduction, notions about the relationship between human
beings and nature, and ideas about people's rights and ethical responsibilities in situations involving risk (Palm, 1990).

In the U.S., individualism and the sanctity of private property are
important cultural values. These values have helped shape the laissez-
faire, persuasion-oriented approach that is generally taken to encouraging the adoption of hazard-reduction measures. While in the interest of
protecting the public's health and safety the government does exercise
some control over what individuals do with their property (for example,
by enacting building codes and land use measures), that control does not
extend to other types of mitigation and preparedness measures that could
effectively reduce disaster losses, such as mandating mitigation measures
for single-family dwellings. This same respect for property (and for the
right to accumulate profit) underlies resistance to measures—including
stricter preparedness and response requirements—that are defined as
overly burdensome by business owners. For example, proposed legislation for measures as diverse as chemical risk assessment, radiological
emergency preparedness, and structural seismic hazard abatement have
been resisted on the argument that their implementation would cause the
economic ruin of affected businesses.

In this same vein, the value placed on individualism in U.S. society
also shapes preferences for certain types of loss reduction strategies. For
example, in attempting to enhance household preparedness and encourage appropriate response behaviors in the U.S., the most common approach is to focus on individual households. In the case of earthquakes,
educational brochures are sent to people's homes along with their Sunday

newspapers, or school children are given materials to take home and share with family members. Far less attention is given to collectively-focused strategies to enhance the disaster-readiness of entire neighborhoods, such as by supporting existing or emergent groups that are already performing similar functions or whose activities could be directed toward preparedness activities.

A belief in the efficacy of technology is another important cultural value that guides U.S. hazard management policy. For years, Americans have acted as if risks can be completely overcome by massive engineered works such as the dams and levees used to control floods and the aqueducts employed to reduce drought vulnerability. In fact, as is discussed in more detail in Chapter Seven, American reliance on engineering "solutions" has been a major factor contributing to the nation's escalating disaster losses.

Despite the American focus on individualism, altruism is a very strong cultural force that also shapes the way disasters are handled. Throughout its history the U.S. has had a strong tradition of volunteer behavior and community involvement, and this altruistic orientation carries over and is amplified in disaster situations. Community residents engage in pro-social behavior on a large scale during an emergency period. Families and neighbors care for one another and donations pour into affected communities (Neal, 1994). Private property is still respected; looting rarely occurs. When it does, it typically involves items of little value (e.g., items picked up as souvenirs by sightseers), and the perpetrators are usually from outside the community. Crime rates generally decline following disaster impact. If illegal behavior does take place—such as breaking and entering for search and rescue or speeding to transport victims—that behavior is redefined by community residents as appropriate in light of the urgent needs of disaster victims.

These are not new findings, but ones that persist (Perry, Hawkins, and Neal, 1983; Neal et al., 1988). In research on all types of disasters, as noted by Goltz, Russell, and Bourque (1992: 45), "there emerges a central theme, that individual and collective behavior is controlled, rational, and adaptive in contrast to popular stereotypes which suggest breakdown and personal disorganization." The increase in altruistic behavior that accompanies disasters also means that victims themselves become valuable resources in preparedness and response efforts (O'Brien and Mileti, 1992).

Cultural expectations also prepare specific occupational groups, such as firefighters, police, paramedics, and other crisis workers to put their own individual self-interest aside when a disaster occurs. Moreover, even members of occupational groups that lack specialized training often show comparable levels of altruistic behavior in disaster situations. During the 1977 Beverly Hills Supper Club fire, for example, cooks and waiters led potential victims to safety and returned into smoke-filled rooms to lead others through back exits (Johnston and Johnson, 1989). Helping behavior was a common mode of response during the 1979 Who Concert "crowd crush" in which 13 people died. Despite the deadly crowding, crowd members assisted each other as much as they could and maintained a relatively functional division of labor (Johnson, 1987).

While post-disaster volunteering and prosocial behavior have been observed not only in the U.S. but also across many different societal settings (see, for example, our discussion on emergent groups in Chapter Three), cross-cultural variations do exist. Following the 1995 Kobe earthquake, which killed an estimated 6,000 people and injured 30,000, many emergent groups formed to assist disaster victims, and organized volunteering took place on a very large scale. As many as 1.3 million people took part in the massive volunteer effort that developed in the days and weeks following that earthquake. Students traveled to the disaster area from around the country, and volunteer groups provided many different kinds of services, from preparing and distributing meals to giving free haircuts. A massive public response of this kind would not have been considered at all unusual in the U.S. However, spontaneous help-giving had not been common in peacetime emergencies in Japan, and the fact that it did occur following the Kobe disaster has become a topic for study among Japanese researchers.

The absence of large-scale volunteerism in Japanese society—both in disaster and non-disaster times—appears to have both cultural and structural sources. Culturally, members of Japanese society tend to feel a much greater sense of social obligation to their families and to secondary groups to which they belong, such as schools and employers, than they do to strangers. Volunteerism, which had had a long history in Japan, declined after World War II as the state and its large bureaucracy increasingly took on functions that had previously been performed in the non-governmental sector. Participation in voluntary groups during non-disaster times is significantly lower in Japan than it is in the U.S. (*The Economist*, 1997). A number of citizen groups that emerged in the aftermath of the Kobe event have continued to exist as organizations, and many provided volunteers to aid in the response to the large oil spill that occurred in the Sea of Japan in early 1997. (For more detailed discussions on volunteers in the Kobe earthquake see Atsumi et al., 1996; Tierney and Goltz, 1997.)

Very little research has attempted to address the social and social-psychological bases of altruism in disasters. Russell and Mentzel (1990), who conducted an experimental study in which 161 Canadian students decided on how much aid to render in 20 different disaster scenarios, found that donations of aid depended upon several factors. First, conceptions of culpability framed their responses; students rendered less aid if they viewed victims as being at fault in some way, for example, by showing poor judgment and putting themselves at risk. Similarly, if an institution or organization was at fault (for example, by not giving adequate warning to victims) the donors gave less aid. Sympathy was related significantly to aid-giving, but only for female donors. Gender was a factor in another way: women tended to give greater amounts of money in aid than men did. The study's authors concluded that the attribution-affect-action effect noticed in routine situations also occurs in disasters. That is, people make attributions about responsibility, respond with a culturally-appropriate emotion, and then act by providing donations.

Dynes (1994) has argued that particular social conditions set the stage for the emergence of altruistic norms. First, a new definition of the situation must be constructed in which potential responders believe that victims are overwhelmed and require a collective response. The media play a particularly important role in providing information to those who are constructing definitions of whether and how to act. At times, the media depict victims as overwhelmed, resulting in a massive response that can inundate disaster-stricken areas (Dynes, 1994; Neal, 1994).

A second condition required for the emergence of situational altruism is linked to changes in normative patterns that arise out of disruptions in everyday-life routines (Dynes, 1994). When disasters force alterations in daily activities and routine patterns of behavior, altruistic responses become more likely. Related social-structural changes constitute the third condition for the emergence of helping behavior on a large scale. When normal routines are upset, people are more structurally available for involvement in organized pro-social behavior. Thus, "mass assaults" by volunteers can be expected in severe and highly disruptive disasters, as illustrated by the 1985 Mexico City and 1989 Loma Prieta earthquakes, as well as by other major disaster events such as the Kobe earthquake.

The collective action that emerges in the context of disasters typically has its origins in pre-existing social groupings, rather than the sudden mobilization of previously isolated individuals. Existing organizations thus make a major contribution to the expansion of human resource availability that occurs during disasters—for example, through extending work hours, switching to double shifts, and taking on volunteers. This type of altruistic emergence peaks during the emergency period. However, some groups that emerge to deal with response-related issues may become institutionalized.

Disaster-induced altruism can have disadvantages, however. Disaster researchers have long been aware that post-disaster convergence can create major management problems and that donations can overburden a disaster-stricken community. Research consistently finds that many donated goods are worse than useless, creating a double disaster for the community residents and organizations that have to deal with them and diverting attention from the real needs of victims (Dynes, 1994; Neal, 1994). After the 1989 Loma Prieta earthquake, for example, health care organizations in Watsonville had to spend personnel and volunteer time setting up a distribution warehouse, and the city of Watsonville had to figure out what to do with planeloads of contributions that were dropped off without warning. After Hurricane Hugo in 1989, the Salvation Army turned tons of donated clothing into a local rag-making factory for later sale. Following the 1995 Kobe earthquake, donors in the U.S. insisted on

Sorting through and delivering donated goods is labor intensive
— and many donations aren't needed.

sending bottled water across the Pacific in huge aircraft at great expense—as if Japan was incapable of providing this most basic resource to its people. Medical supplies and vaccines were delivered with great fanfare (and public outcry when some were refused), ignoring the fact that Osaka, only a few miles from the impact area, is one of Japan's major pharmaceutical centers.

Study after study shows that those attempting to provide aid are often unaware of the real needs of disaster victims. One report, for example, indicated that only 30 percent of the drugs sent to Armenia after the 1988 earthquake were immediately useful and that the remainder were too poorly labeled to be of use, not suitable, or expired or frozen when they arrived (International Federation of Red Cross and Red Crescent Societies, 1993). Carter (1991: 117) gives other examples:

> In one recorded case, a large supply of yellow bikinis was sent to refugees trying to subsist in semi-arctic conditions. In other cases, supplies of high-heeled shoes were sent to victims who were never likely to wear them. In a third case, a well-meaning overseas community collected a huge amount of fruit and had it flown by chartered aircraft to a neighboring country. On arrival, the fruits had to be destroyed because of the danger of introducing fruit-fly and thus risking the future of indigenous crops.

Dynes (1994) has noted that situational altruism can have other unanticipated negative effects, for example by fostering victim dependency, which can in turn be used to justify organizational existence. Despite such problems:

> The more important lesson, however, is often lost that situational altruism provides the resources, human and material, to create an effective emergency response. While it has its inefficiencies, it is usually more effective than the rational solutions which are currently offered as improvements (Dynes, 1994: 16).

Dominant cultural assumptions can also disadvantage some groups when disasters strike. In all societies, the interests of socially and economically privileged groups are embedded in mainstream cultural practices, which in turn become ingrained in the ways in which organizations and institutions operate. The literature on disasters contains numerous examples of the ways in which these hegemonic ideas shape the activities of preparedness and response organizations, resulting in the failure to take cultural differences into account in the delivery of disaster-related services. In disasters, the practices of crisis relevant organizations reflect prevailing social hierarchies and the differential value placed on different groups, subcultures, and lifestyles. These influences are becoming more evident as researchers have turned their attention to the ways in which social and cultural diversity shape disaster experiences and on recent major disaster events that have affected highly diverse communities (Phillips, Garza, and Neal, 1994; Bolin and Stanford, 1998; Peacock, Morrow, and Gladwin, 1997; Enarson and Morrow, 1998). As these studies show, because services are geared toward the needs of the dominant majority, minority disaster victims in shelters may be given food that is very different from what they ordinarily eat, non-native English speakers may be required to fill out extensive forms in English, and immigrant victims may be justly fearful of seeking needed services for which they qualify out of fear of deportation. Programs can fail to accommodate households made up of more than one nuclear family and otherwise fail to recognize cultural variations in living arrangements, such as those involving extended and multi-generational families and "doubling-up" for economic reasons. Yelvington (1997: 109), for example, observed this pattern in the aftermath of Hurricane Andrew in Florida:

> Advocacy groups complained that the biggest obstacle was the FEMA rule that stated only heads of household—as defined by FEMA—were eligible to receive trailers and other forms of aid . . . In households shared by three or four families, FEMA either denied the entire household money or awarded assistance to only one family . . . given the complex ethnic, cultural, and class makeup of South Dade [County], official policies often did not match the realities of victims' lives.

Service providers may employ other criteria of aid-worthiness that reflect the values of the dominant culture. For example, in recent disasters, researchers have documented agency's efforts to distinguish between the "deserving" homeless—that is, those needing shelter because they lost their homes due to disaster impact—and the "undeserving," "pre-disaster" homeless and to take steps to ensure that the two groups are segregated from one another when shelter is provided (Bolin and Stanford, 1993; Phillips, 1996, 1998). As noted in the previous chapter using an example from Hurricane Andrew, ethnic and income-based differences in the provision of governmental disaster assistance have also been documented at the community level, again suggesting that culturally-related stereotypes influence judgments about who deserves disaster aid.

Even though this idea has been given little explicit emphasis in the literature on U.S. disasters, it is also evident that cultural practices are a source of knowledge, sustenance, and resiliency for people who must cope under conditions of disaster threat and impact. Oliver-Smith (1986,

1994), for example, demonstrated how the Andean peoples of the pre-Columbian period learned to adjust to earthquake hazards, settling the land and designing and building structures to protect against earthquake-related death and injury. These pre-Columbian cultures also developed elaborate storehouse systems that provided for people's needs during times of disaster and environmental hardship. The understandings and practices that were incorporated into disaster subcultures represented learned ways of coping with repeated exposure to particular hazards. Residents of hazard-prone areas drew upon a stock of knowledge that enabled them to live with threats, accurately interpret environmental cues, and take appropriate action when disaster struck.

Examples from other recent disasters show how cultural practices and local knowledge help people respond more effectively when disaster strikes. In Guadalajara, Mexico, community residents who successfully searched for and rescued their neighbors and loved ones when a massive gas explosion rocked their neighborhood in 1992 did so not because they had access to sophisticated technologies or equipment, but rather because they had an understanding of daily life routines that enabled them to discern who would need help and where those individuals would likely be found (Aguirre et al., 1995). Similarly, when Mexican and Central American immigrants in California leave their homes and shelter outdoors when an earthquake strikes, even when their homes are judged by inspectors to be safe to inhabit, they may create problems for official response agencies, but their actions are based on their own culturally-based notions of how to ensure life safety in earthquake situations. Even buildings without significant structural damage have been known to collapse in large aftershocks, and the desire to remain outdoors represents an ingrained cultural aversion to that risk.

Critical and alternative cultural orientations also challenge the dominant culture's assumptions about hazards and their management. Eco-feminism, deep ecology, and new environmental movements, for example, offer perspectives on the relationship between people and the natural world that differ radically from those of mainstream culture (Devall and Sessions, 1985; Warren, 1994; Murphy, 1994; Gaard, 1998). These alternative views call into question many assumptions about society/environment relationships that are taken for granted in the U.S. and other industrialized countries, including those that underlie our approaches to managing hazards. For example, they reject the idea that nature exists solely to benefit human beings and the notion that large-scale public works and advanced technologies should be employed to control natural processes. Applied to hazards, such critiques also lead logically to questions about whether higher levels of economic development and affluence invariably promote safety and whether government agencies and large bureaucratic organizations, rather than community residents themselves, are best able to provide solutions for hazard-related problems. Although these perspectives have not yet had a discernible influence on the hazard policy process, they have begun to have an impact on public discourse on environmental hazards.

ECONOMIC FACTORS

Although past research on loss reduction has tended to gloss over economic issues, economic forces clearly play a role in determining both disaster losses and what is done to deal with them. For example, as noted in Chapter Five in our discussions on the role of income and other status-related factors in the adoption of self-protective measures, economic resources are related to disaster vulnerability at the household level. No one would argue that the rich always escape the effects of disasters. After all, Malibu, the Oakland Hills, and the East Coast Barrier Islands are meccas for the rich, and they have all suffered serious disasters in the last few years. However, other things being equal, well-off people also tend to fare better during disasters because they are better able to shield themselves from damage and disruption and better able to recover when they do suffer losses.

It is equally clear that organizational, community, and societal loss-reduction measures—including those involving emergency preparedness and response—are influenced by broader economic (or, more accurately, political-economic) forces. Cross-national comparisons make it clear that countries with higher per capita incomes tend to be safer, whether the focus is on general health and safety or on disasters. Deaths resulting from natural disasters in less-developed countries exceed by many orders of magnitude those in the developed world. The 1976 Guatemala earthquake, for example, killed 22,000 people, and the Tangshan earthquake that same year killed an astounding 240,000. At least 10,000 died in the 1985 Mexico City quake, and the death toll in the 1988 Armenian earthquake is estimated at around 25,000. In contrast, comparable-sized earthquakes that took place in the U.S. during roughly the same time period resulted in far fewer fatalities. The 1971 San Fernando earthquake killed 64, the 1989 Loma Prieta event killed 67, and the Northridge earthquake of 1994 resulted in 33 deaths.

Indeed, these kinds of disparities are found within countries as well. For example, the poor bore the brunt of the earthquake's impacts in Guatemala while the rich, whose homes were constructed to resist seismic forces, suffered relatively few fatalities. Differences in the vulnerability of the rich and poor were so stark that the earthquake was referred to by some as a "class-quake" (Blaikie et al., 1994). None of these earthquake-stricken societies lacked the technical knowledge necessary to make the built environment more earthquake-resistant. The problem was that this knowledge was applied only selectively—again illustrating the importance of economic factors.

This is not to say that rich societies are equally safe for everyone, or that they do not produce risks. As noted above, economic resources are associated with differential access to disaster protection both across and within societies, and vulnerability is stratified even in the most affluent countries. The better-off nations of the world have also been responsible for introducing technologies such as nuclear power that create new and potentially deadly hazards. And a predominant pattern in the late-twentieth century is the transfer of risky technologies and products from more-developed to less-developed societies, making life safer for those living in better-off countries but more risky for others (see, for example, analyses by Frey, 1995, on the risks that the international trade in pesticides pose for less developed countries).

Related to this point, the operation of the global political economy, which involves power-dependency relationship between rich and poor countries, contributes heavily to disaster vulnerability (Hewitt, 1983; Blaikie et al., 1994; Tierney, 1999). The differences that exist between countries in the center and on the periphery of the world economic system include significant differences in their degree of exposure to disaster losses. We will return to this idea in Chapter Seven when we discuss disasters and issues of sustainability.

While a thorough consideration of the role of economics in disaster vulnerability and loss-reduction is beyond the scope of this book, based on the current disaster literature we can make a number of observations on how economic factors operate in areas involving hazard reduction and risk. The first is that until recently the potential economic impacts of disasters have not been seen as very significant in this country, either by researchers, the general public, or policymakers. Research that has been conducted to date on U.S. disasters suggests that they have few discernable negative economic consequences at the community and regional levels (Friesema et al., 1979; Rossi, Wright, and Weber-Burdin, 1982;

Cochrane and Schmehl, 1993). Economic analyses also suggest that disasters can have positive economic impacts regionally, due to the stimulus provided by reconstruction, although they are also accompanied by a slight impact nationally in the other direction (Cochrane, 1997).

Compared with the massive size of the U.S. economy and the overall value of the built environment, the effects of disasters—even very large ones—have generally been minor. Neighborhoods and smaller communities may suffer terribly when disaster strikes, but economic activity at larger levels of aggregation has proven to be quite resilient in the face of disasters. Similarly, disasters may cause some businesses to fail, but those businesses typically are replaced by others, so that aggregate effects are minimal. The fact that U.S. communities have recovered from even the largest disasters (the 1906 San Francisco and 1964 Alaskan earthquakes are cases in point) supports the contention that disasters do not have discernable economic effects in the long run.

Both cultural and institutional practices reinforce the idea that the economic aspects of disasters merit little concern. Although there is some evidence that this view is changing (see our discussions above on social-constructionism and disasters), most Americans consider natural disasters to be "acts of God" or "acts of nature" whose effects are more or less random and accidental. Because disaster impacts are defined as involving fate rather than choice, it follows that victims are blameless and should be assisted in recovering from their losses by the government, private sector organizations like the Red Cross, and insurance. Americans respond to the plight of disaster victims both here and abroad with great sympathy and generosity. The public, in other words, frames the disaster problem in human, rather than economic terms.

These cultural beliefs have a parallel in the ways in which government programs and markets respond to hazards. In general, governmental hazard management programs have not attempted to discriminate between prudent and imprudent households, businesses, or communities. The National Flood Insurance Program is a possible exception, because it does attempt to distinguish between communities that institute good flood-control policies and those that do not. However, there is evidence to suggest that it has not worked as intended and that it may even lead to higher flood losses. Since the program has not been systematically evaluated, we lack a good understanding of its actual impacts.

Similarly, until recently insurers have not attempted to reward loss reduction activities through their rate structures. Real estate prices are not sensitive to hazards, as they are to school quality, environmental quality, and amenities such as attractive views, indicating that location in a risky area doesn't detract from the value of property (see, for example, Palm, 1990, on real estate markets and earthquake hazards). In short, until recently there has been a de facto assumption on the part of both the public and key institutional actors that the losses associated with disasters are acceptable, at least from an economic point of view.

Despite this seeming inattentiveness to the economic dimension of hazards, the disaster literature contains many discussions on the ways in which economic forces, incentives, and disincentives help bring about disaster losses. Obviously, development pressures are implicated in escalating disaster losses, as population density has increased in high-risk areas. Land-use regulations are among the most potentially effective tools for reducing losses, but they are also among the most difficult to implement, because powerful economic interests typically oppose controls on land development. The same pattern can be observed with stricter building codes and special hazard-reduction ordinances; opponents of such measures include development interests and landlords trying to avoid the costs associated with compliance (Alesch and Petak, 1986).

Various government reports also have noted that neither the U.S. policy system nor the private sector offer economic or fiscal incentives that would encourage the adoption of loss-reduction measures by households, organizations, and communities. Government at all levels appears reluctant to employ potentially powerful economic tools it has at its disposal to contain future disaster losses. For example, a report to Congress by FEMA, dealing with barriers to earthquake hazard mitigation (1993), noted that if the federal government were to extend Executive Order 12699, which regulates levels of seismic safety in government-owned and -leased properties to include similar requirements for properties purchased with *federally-assisted* mortgages issued by FDIC-backed financial institutions, the result could well be a significant improvement in the seismic safety of the built environment. One of the key conclusions of the report (1993: 18) involved economic incentives the federal government could use, but to date has not:

> Broad national requirements for receipt of federal support would oblige subfederal governments to assign high priority to the issue [of seismic safety] and, at the same time, insulate local officials from the kinds of criticism and opposition they currently face if they try to act on their own to reduce hazards . . . If the federal government were to adopt this strategy, it would demonstrate that the government is not simply imposing rules and regulations but is enabling society to avoid earthquake damage, while at the same time rewarding those who engage in mitigation.

The same is the case for the private sector. Existing hazard insurance merely redistributes risks rather than reduces them because insurers have generally been uninterested in encouraging mitigation. Judging from their lack of involvement in promoting hazard reduction, other important financial sector institutions such as the banking industry are also less than eager to support strategies that would encourage higher levels of disaster protection.

In short, in the disaster area the policy system permits vulnerability to escalate while failing to provide sufficient rewards for risk avoidance. Choices that have an impact on how much risk is assumed have a very strong economic component. As noted above, when households and businesses decide on loss-reduction measures, they tend to favor those that are easy and inexpensive to undertake. A business owner is much more likely to keep a first-aid kit on hand than to have a back-up generator or to develop a business continuity/recovery plan because of the financial and time investment involved in taking those steps. Owners are also more likely to undertake low-cost preparedness measures than to have their buildings structurally assessed or strengthened to resist disaster damage. In a system that fails to provide tax relief or low-interest loans for structural upgrading, such decisions may well be economically rational, particularly with respect to perceived low-probability events.

Likewise, the mix of loss-reduction strategies communities adopt is also highly influenced by economic considerations. In fact, a case can be made that preparing for disasters and responding when they occur are emphasized over other long-term mitigative actions in most U.S. communities precisely because they are less expensive and easier to sell politically. Different political and economic stakeholders can agree that at minimum government should be able to save lives and deal with property damage should a disaster actually strike—and indeed, government is mandated to do so—even if these same stakeholders find it impossible to reach a consensus on larger financial investments that would reduce the overall threat to life, safety, and property. Seen in this context, emergency preparedness and response programs are bare-bones investments that substitute for more difficult choices.

TECHNOLOGICAL CHANGE

The technologies that are available to societies, communities, public and private organizations, households, and other social units clearly have a major impact on the ways in which disasters are managed. In industrialized societies like the U.S., the ability to plan for and respond to disasters increasingly rests on technologies that can help identify hazards, detect impending disasters, warn the public of immanent threats, and facilitate communication among responding organizations, levels of government, and residents of disaster-stricken areas. Improvements in technology have made it possible to perform many disaster-related tasks more efficiently and effectively, protecting life and property and reducing disaster losses. Evidence of the ways in which technology contributes to disaster preparedness and response is abundant, from the satellite imagery that tracks hurricanes before landfall, to technologies that detect oncoming tsunamis and issue warning signals, to ultra-sensitive listening devices and sensors that can detect the presence of human beings trapped in collapsed structures. The pace of technological change in the disaster field has increased in the past few years, driven both by the information revolution and by the transfer of technologies previously used in military and defense applications to the management of hazards. At the same time, although technology has much to offer, it would be a mistake to adopt untested technologies uncritically. Since a complete review of all the ways in which technology affects disaster preparedness and response is beyond the scope of this volume, our discussion will focus briefly on illustrating how selected technological developments and trends have changed the ways in which American society attempts to manage hazards.

Since the time of the first research assessment, no technological change has had a greater impact in the hazards management area than advances in computers and information technology (IT). Developments in IT offer the promise of a host of hazard-management improvements: more systematic and timely hazard and vulnerability assessments; improved warning systems; the capacity for launching a more efficient and effective organizational response; greater ability to anticipate response-related problems; enhanced organizational capacity; improved ability to track resources as they are mobilized and to detect disaster-related problems as they develop; and expanded data-collection and crisis management capability. A number of tools are now at least theoretically available to emergency officials for use in disaster situations, although questions remain about their implementation and their usefulness in actual disaster situations. (For good discussions of IT applications in both national and global emergency management see Comfort, 1993; Disaster Information Task Force, 1997.)

Scholarly interest in emergency management applications for computer technologies began to take shape with the publication of the edited volume, *Terminal Disasters* (Marston, 1986). Papers in that collection reviewed various ways computers can be used in emergency management, including their use for information management and crisis decision making; modeling natural processes, such as storm surges; identifying populations at risk and simulating the movement of people in disaster situations, including evacuation behavior; and forecasting damage to the built environment to improve planning and response. Several years later, Drabek (1991c) studied emergency management offices in four states and in 12 communities in those states. That research indicated that states and localities differed in the manner in which they adopted and implemented computer technologies. Barriers to implementation, which were substantial, included issues related to staffing needs, funding, and hardware, software, and data base compatibility. Computer use was found to have had an impact on agency structure—for example, in staffing patterns and staff qualifications—as well as on organizational culture and interorganizational relationships. Based on his research, Drabek concluded that the changes that would be brought about by increased adoption of computer technologies would be profound and far-reaching. With respect to crisis management, for example, he argued that (1991c: 182) "[i]ncreased microcomputer implementation by local and state emergency management agencies could do more to enhance disaster response capacity than any other single change." The use of computer data bases, modeling tools, and other information technologies has expanded greatly since *Terminal Disasters* and the Drabek study were published. In the sections that follow, we briefly discuss two broad but overlapping areas in which computer applications have been used extensively: vulnerability assessment and response-related decision making.

Information Technology, Vulnerability, and Crisis Management

Advances in computer technology have given scientists and emergency managers both greatly enhanced computational and modeling capability and the ability to manipulate large amounts of data in order to anticipate disaster-related problems. One illustration is the computer applications that have been employed for weather forecasting, plotting where and when storms will strike, and projecting various kinds of impacts they will have in populated areas. One such analytic technique, the "Sea, Lake, and Overland Surges from Hurricanes" (SLOSH) model formulated by the National Oceanic and Atmospheric Administration, is currently used extensively in hurricane response management. SLOSH

*Hyogo Prefecture in Japan equipped its EOC with new
technology after the 1995 Kobe Earthquake. (Kathleen J. Tierney)*

was originally developed primarily as a meteorologic tool for identifying geographic areas that would be affected by hurricane storm surges. (The forerunner of SLOSH was a model called SPLASH, which modeled storm surges along open coastlines. That model lacked the complexity of SLOSH and, unlike SLOSH, it could not predict the size of inland storm surges.) When combined with population data and a strategy for disseminating the information on storm surge and flooding potential to the public, it has also become a resource for hurricane preparedness and evacuation planning and response. SLOSH was first used for hurricane evacuation planning in the Tampa Bay, Florida, area in 1980; since that time, models have been developed for many hurricane-prone population centers around the country.

Many computer-aided approaches to disaster planning and management employ geographic information systems (GIS) (for a good overview on GIS applications in disaster management and research see Dash, 1997). The application of GIS technology to disasters is very appropriate, since GIS makes it possible to integrate geographic, spatial, or locational data, such as information on the scope of a disaster's impact, with other types of data, such as information on the characteristics of the built environment and of the affected population. GIS techniques have been used both for assessing pre-event vulnerability and for developing post-event response and early recovery strategies. With respect to earthquakes, for example, data on seismic hazards (e.g., ground shaking and liquefaction) have been combined with data on the built environment and population characteristics in earthquake-prone areas to project damage, losses, and population impacts.

The PEPPER (Pre-Earthquake Planning for Post-Earthquake Rebuilding) project (Spangle, 1987) was the first major effort to apply GIS-based techniques to a community—in that case, the city of Los Angeles—to assess vulnerability, identify which areas in the community would be hardest-hit in an earthquake and where recovery needs would be greatest, and begin estimating effects on the population such as damage and loss of residential property. Since the time of that first project, the use of GIS in disaster loss estimation has expanded enormously. Beginning in the early 1990s, with funding from FEMA the National Institute of Building Sciences oversaw the development of a GIS-based package of earthquake hazard and vulnerability assessment software for use by local governments. The system, called HAZUS, models both direct damage to buildings and lifelines and a range of ancillary and indirect losses, including fire following earthquake, homelessness due to building damage, and hazardous materials releases. HAZUS was intended as an "off-the-shelf" software package that could be distributed to earthquake-prone communities for use with locally-available data on the built environment. Pilot-testing of the HAZUS methodology indicated that it produced more accurate estimates in areas of high seismic activity than in low-seismicity parts of the U.S., and the model is undergoing refinement. Initially developed for earthquake hazards, HAZUS is now being extended for use with other hazards, such as floods and high winds.

The first large-scale attempt to use GIS in a major disaster was made when Hurricane Andrew struck in 1992. After that event, FEMA began using GIS tools to provide needed information to federal, state, and local agencies. Although it took many weeks to develop the necessary databases, GIS software and datasets were eventually used to document, analyze, and track both disaster impacts and recovery-related activities (Dash, 1997).

Following the Northridge earthquake, a GIS-based system called EPEDAT (Early Post-Earthquake Damage Assessment Tool) was used in the first hours after the earthquake to estimate potential dollar losses due to earthquake damage. Those estimates then served as the basis for the state of California's request for federal disaster assistance. EPEDAT could calculate aggregate damage estimates for the impact region because the system was able to relate extensive data on the built environment (e.g., building construction types and sizes) to data on the shaking intensities associated with the earthquake. Projections on damage derived from EPEDAT calculations as well as damage data obtained from building and safety departments were also used to identify areas where the need for disaster assistance would be greatest, and this information helped officials decide on where to locate disaster application centers. Because of their ability to relate damage data to population data available from the U.S. Census Bureau, emergency management agencies knew which segments of the population were hardest hit and were better able to anticipate their needs. For example, they determined that in certain damage areas many of the victims would be non-English speaking, and they were able to recruit translators to work at disaster assistance centers.

During the 1990s, GIS technologies have become well established as tools for aiding mitigation, preparedness, response, and recovery activities. The adoption of GIS has been driven in part by the sustained marketing efforts of GIS companies such as Environmental Systems Research Institute (ESRI), developers and vendors of ARC/INFO, ARCVIEW, and a range of other GIS-based tools. GIS software vendors recognized early on that, because of their spatially-distributed and at the same time socially-complex nature, disasters lend themselves very well to the kinds of analytic applications GIS systems offer.

Even though GIS technology is widely available, and even though its potential as a disaster management technology is widely recognized, that potential is often unrealized, for various reasons. At the local level, GIS experts tend to be concentrated in planning departments rather than in local emergency management agencies, which may lack the personnel and training to take advantage of what GIS can offer. Additionally, as Dash (1997) points out, GIS analyses are only as good as the data on which they are based. Since developing complete and accurate datasets, such as detailed records on the characteristics of the built environment, is expensive and labor intensive, many local communities may simply be unable to make the investment. GIS systems themselves are expensive to operate and maintain, particularly if personnel costs are taken into account. Substantial computational capacity is needed to run very large data sets quickly enough to allow analytic results to be useful for disaster response management. While some very large cities may have access to highly sophisticated computer equipment, most communities do not. Attaining higher levels of technological sophistication may be financially difficult for many crisis-management organizations and communities, particularly smaller ones.

Currently there is also a tendency for GIS technologies to be used more for descriptive and representational purposes than for research and policymaking. Critics argue that GIS is often employed mainly to produce "pretty maps" which are then used to brief politicians and the media. In other words, GIS is used to show *what* happened in a disaster situation, but not *why* it happened or what can be done to avoid future problems. Its analytic power is rarely used to develop a deeper understanding of disaster processes and impacts or to serve as a basis for decision making. That situation may improve as the technology becomes more widely used and accepted, as more GIS-trained personnel are employed by hazard-related agencies, and as key actors come to recognize how GIS can be used to explain and analyze rather than merely to describe, hazard-related phenomena.

Since the early 1990s, enthusiasm has also been steadily building for the transfer of a number of advanced technologies, including remote-sensing and information-processing technologies formerly used for other purposes, to the hazards area. These technologies include global positioning satellite systems (GPS), synthetic aperture radar systems (SAR), and high-performance computing. With the dissolution of the Soviet Union and the end of the Cold War, U.S. agencies such as the Department of Defense and the Central Intelligence Agency—which had previously used highly advanced technologies mainly for purposes such as national defense and surveillance—have been showing an increasing interest in monitoring hazards and producing various types of data and information for disaster management purposes (for good discussions of this trend see Pace, O'Connell, and Lachman, 1997; Quarantelli, 1998b). A major initiative in this area, the Global Disaster Information Network (GDIN), would focus on both nationwide and cross-national sharing of data and information—including sanitized intelligence information—in the management of disasters (for a description of the planned system see Disaster Information Task Force, 1997). Other researchers and practitioners have also proposed a Global Emergency and Risk Management System (GERMS) that relies extensively on advanced technologies (Eguchi et al., 1998). Clearly an enormous amount of data, software, and advanced communications technologies currently exists that have potential applicability for loss reduction. What remains to be seen is how such technologies will actually be used by decision makers during disaster situations and what their impacts will be.

The Internet, another technological innovation whose societal effects are already far-reaching, has also begun to have a major impact on preparedness and response activities (see, for example, Anderson, 1995; Botterel, 1995–96; Gruntfest and Weber, 1998). Enormous amounts of information on all aspects of hazards and disasters are now available from both official and unofficial sources on the World Wide Web. The Internet is increasingly being viewed as the ideal mechanism both for disseminating information to the public and for coordinating organizational and community activities when disaster strikes. Clearly, its use and efficacy in actual disaster situations are topics that warrant in-depth study. For example, to mention just one topic—the provision of hazard- and disaster-related information to the public—it is clear that Internet-based communications media are making a wider array of information available than ever before on the hazards communities face, on how to mitigate and prepare, and on the impacts of disasters when they do occur. Unfortunately, however, the Web could also be a vehicle for the dissemination of incorrect information, poor guidance, rumors, and disaster myths.

Another issue that has not yet been addressed systematically by researchers involves the capability and readiness of emergency management agencies to employ new technologies. Currently, much more emphasis is being placed on developing technology-based emergency management solutions than on assessing emergency management needs or exploring how to integrate complex technologies into existing emergency management organizations.

From the time the use of computers and other technological tools began to be studied in the early 1980s, researchers have observed that wide variation exists nationwide in technology adoption and application (Marston, 1986; Drabek, 1991c). Although we know that the use of new technologies is now widespread and increasing exponentially within the emergency management field, so little research has been undertaken on the diffusion of technology that we currently have no way of accurately gauging how various technologies are being used, by whom, and to what effect. New and emerging technologies offer great promise. Whether they can be implemented as planned in actual disaster situations is a topic for further research.

There are many reasons to remain skeptical about the idea that technology will provide a panacea for emergency management problems. Although software packages and decision-support tools are now widely available, we have little information on how they are actually being used to manage hazards. Lack of organizational capacity has historically impeded preparedness and response efforts. Disasters are generally not a high priority for most governmental units, and disaster-related needs rarely receive the resources they warrant. Merely providing technology will not change that situation. Strategies that stress reliance on computers in disaster situations assume that technological knowledge is widely shared among potential users—an assumption that has not been empirically explored. More-intensive use of new technologies may well serve to reinforce the social inequities we have highlighted throughout this book. Poor people and members of minority groups tend to have significantly less access to IT than the better off and members of the white majority. These "technology have-not" groups, which are already at a disadvantage in so many other ways, could also be left behind in an emergency technology revolution that fails to address their needs.

In a series of insightful analyses, Quarantelli has reflected on emergency management's national and global "information/communication revolution" from the point of view of research on hazards, on technology transfer and diffusion of innovations, and on socio-technical systems (see Quarantelli, 1997, 1998b, 1998c). While recognizing that the application of new technologies to disaster-related problems can improve hazard management in many ways, Quarantelli also identified a number of potential negative effects and unexamined assumptions associated with their use. Among the problems noted are that introducing advanced technologies will almost certainly widen the gap between the rich and poor in U.S. society, as well as between industrialized nations and the developing world; that proposed technological fixes can become ends in themselves, driving organizational decisions and priorities rather than the other way around; and that the communications revolution can result in information overload and the dissemination of incorrect and outdated information as well as in accurate guidance. Particularly in large-scale events, the availability of a range of communications media can result in a convergence of information that parallels the physical convergence that has long been observed in disaster situations. Overreliance on new technologies may actually undermine the ability of organizations to learn from their own mistakes. And finally, Quarantelli also stressed the need for recognizing that management of hazards is fundamentally social in nature and not something that can be achieved strictly through technological upgrading.

In our fascination with the promise of technology, we should not lose sight of the fact that many proven ways of dealing with disaster-related problems are and will remain decidedly low-tech. The notion that technology will save us is, in other words, as invalid for hazard management as it is for social life in general. Even with the most advanced disaster warning technologies, people still need to confirm warnings by communicating with their friends and neighbors, and garbled and confusing warning messages will still impede action, even if those messages are conveyed with lightning speed. Even in an era when high-tech search and rescue equipment is commonly available at disaster sites, survivors still tend overwhelmingly to be rescued by their family members, friends, and neighbors, working with their own hands. While technology can help in many ways, a great deal of the work that needs to be done in disasters still involves routine labor on the part of large numbers of people: sandbagging, cleaning up debris, handing out food and water, and providing face-to-face help and advice to victims.

Finally, since many disasters can render useless the very technologies on which our society has come to rely, we should avoid being overconfident that those technologies will be there when we need them. Indeed, as devastating disasters like Hurricane Andrew have illustrated, disaster can often involve the extended loss of even the most taken-for-granted daily needs, such as television, electricity, and air conditioning. A computer that is drenched with water or buried under debris is useless, especially if its power source is unavailable. Rather than believing mistakenly that they will always be able to turn to the Internet or the web for information when disaster strikes, community residents and those who advise them on disaster preparedness should work from the premise that commonly-available technologies may well fail when disaster strikes, and should plan accordingly.

PROFESSIONALIZATION AND KNOWLEDGE TRANSFER

Among the most important changes affecting emergency preparedness and response since the time of the first assessment has been an increasing trend toward professionalization in the field of emergency management. A generation ago, emergency management did not exist as a recognized profession. Individuals were considered qualified to assume the position of civil defense director if they had undergone what were judged to be relevant training experiences in the fire service, the police, or the military. Civil defense directors typically had multiple responsibilities. For example, a fire chief or, less frequently, a police chief might be assigned the title of civil defense director as a collateral duty. Being responsible for community preparedness and response activities was not considered a full-time job, and the skills needed to perform the job were ill-defined.

The role of the emergency manager began to evolve into a profession during the 1970s. Local jurisdictions began increasingly to identify the emergency manager's position as a full-time post. The conception of the role also began to broaden beyond that of "civil defense" and the conduct of immediate post-impact emergency activities. As the notion of comprehensive emergency management gained currency in the late 1970s and early 1980s, the job expanded to include mitigation and recovery as well as preparedness and response.

Over time, it also became increasingly clear that to perform effectively emergency managers need to be more than good planners. They must possess technical knowledge of the hazards facing their communities and must be able to communicate this information effectively to local officials and the general public. They need to have knowledge of the emergency management resources and programs that are available outside their jurisdictions, including programs for providing financial assistance for training and preparedness as well as the emergency response resources that are available through state and federal agencies, industry associations, and professional societies. Once they obtain external resources, they need to manage them effectively. Emergency managers also need to have a good grasp of how governmental systems operate, and they must be politically adroit in mobilizing support for emergency management in their communities (Perry, 1991). In short, the job is now seen as requiring someone who is equally competent as a technical expert, program administrator, and politician.

This process of professionalization has been accompanied by the formation of associations concerned with the training and credentialing of emergency management specialists, the development of publications geared specifically to practitioners, the diffusion of research findings into the practitioner community, the growth and spread of professional meetings and conferences, and other changes indicating that emergency management has emerged as a specialized discipline. This move in the direction of greater professionalization had a very important impact on disaster preparedness and response.

Trends and Influences in the Professionalization of Emergency Management

In the late 1960s and early 1970s, a series of major disasters and a partial easing of Cold War tensions served to direct attention toward the disaster problem. An underlying theme in the development of the profession has been increasing concern about the inadequacies of programs designed to protect the public against disasters and other emergencies (National Academy of Public Administration, 1993).

As noted above, from the time of the first governmental attempts to enact coherent disaster legislation in 1950, federal management efforts were fragmented, subject to almost continuous revision, and ambivalent with respect to goals. In 1979, the National Governors Association (NGA) published a landmark report expressing concern about the lack of a comprehensive national policy to manage emergencies and the dispersion of responsibility for disaster management among numerous federal agencies (National Governors Association, 1979). Among the NGA study's findings were that state programs mirrored the federal government's fragmented approach to the disaster problem and that programs generally lacked an integrated approach to managing hazards—that is, a set of management strategies encompassing mitigation, preparedness, response, and recovery. The report called for federal, state, and local governments to enter into an equal partnership and to adopt a comprehensive approach to emergency management. It also recommended the creation of a federal agency and counterpart state agencies to coordinate emergency management activities. That same year saw the initiation of President Jimmy Carter's reorganizing project, which resulted in the formation of the Federal Emergency Management Agency.

Since that time, programs initiated by the federal government have had a pronounced impact on training and practice in the emergency management field. FEMA sponsors two major training facilities, the Emergency Management Institute (EMI) and the National Fire Academy (NFA), both of which are housed at the National Emergency Training Center in Emmitsburg, Maryland. EMI provides instruction in emergency management for state and local officials, emergency managers, volunteer organization personnel, and practitioners in related fields. Each state emergency management office has a FEMA-funded training officer who coordinates the delivery of federally-funded training programs throughout the state. EMI is matched on the state level by training centers such as the California Specialized Training Institute (CSTI), which is a branch of the Governor's Office of Emergency Services (OES) devoted specifically to improving knowledge and skills in the areas of emergency and disaster management.

Specialized associations have also served as a vehicle for professionalization. The International Association of Emergency Managers (IAEM), formerly the National Coordinating Council on Emergency Man-

agement (NCCEM), was founded in 1952. IAEM is a non-profit association of approximately 1,600 individuals and organizations from the local,
state, and federal levels as well as from the private sector and the military.
The organization has offered a certification program in emergency management since 1993. Organizations such as the Association of Contin-
gency Planners (ACP), whose membership consists primarily of individuals who give emergency management guidance to private-sector entities,
also indicate growing recognition of emergency management as a specialized field. The association of Voluntary Organizations Active in Disasters
(VOAD) is another professional group that provides a forum and organizational infrastructure for nongovernmental organizations whose missions
center on the provision of disaster-related services.

Other developments have contributed to professionalization and
knowledge exchange among hazard researchers and emergency management practitioners. In July of 1983, FEMA and the National Association
of Schools of Public Affairs and Administration (NASPAA) co-sponsored
a conference on Emergency Management in Public Administration at the
National Emergency Training Center as a first step in examining how
emergency management could be incorporated into public administration education. One of the products resulting from that conference was a
special issue of the journal *Public Administration Review* (Petak, 1985)
focusing specifically on emergency management as an emerging field
within public administration. Further adding to the visibility and professionalization of the field, in 1986 the American Society of Public Administration (ASPA) established a section on emergency management.

Since that time, disaster- and emergency-management related courses
and degree programs have also begun being offered at colleges and universities nationwide. A number of institutions of higher education and
various academic specialties now offer emergency management and disaster-related courses. The University of North Texas was the first institution of higher education to offer an independent undergraduate degree
in emergency management. Majors must complete 36 hours in the emergency management field in addition to the university-required core cur-
riculum. The program, which has been in existence since the mid-1980s,
and which typically has about 150 majors at any given time, has con-
ferred more than 300 Bachelor of Science degrees. A number of other
educational institutions have also begun offering courses, areas of specialization, degrees, and certificates in emergency management in both
campus-based and distance-learning formats. Many of these programs
target emergency management practitioners.

During the 1990s, FEMA developed a project on higher education to stimulate the incorporation of hazard-related topics into college and university curricula and to make research-based knowledge more available to those who wish to obtain advanced training in emergency management. As part of that project, FEMA has assisted with the development of college-level course curricula covering a range of fields and topics, including the sociology of disasters, technology and emergency management, and the political and public policy aspects of emergency management. Nearly two dozen other courses are in the process of being developed on such varied topics as hazard mitigation, disaster recovery, and the economics of hazards and disasters. Table 6.1 lists courses that have already been developed with FEMA's support, as well as those that will soon be made available.

While the field of disasters and hazards remains a relatively small research specialty, the past 20 years have seen an impressive proliferation of college- and university-based research, training, and educational

TABLE 6.1 Existing and Planned FEMA Higher Education Project Courses

- Building Disaster Resilient and Sustainable Communities
- Business and Industry Crisis Management
- Disaster Response Operations and Management
- Earthquake Hazard Management and Operations
- Economic Dimensions of Hazards and Disasters
- Emergency Management for the Fire Community
- Emergency Management Principles and Application for Tourism, Hospitality, and Travel Management Industries
- Emergency Management Skills and Principles
- Hazards, Disasters and the U.S. Emergency Management System
- Hazards, Vulnerability and Risk Analysis
- Individual and Community Disaster Education
- Living in a Hazardous Environment
- Political and Policy Basis of Emergency Management
- Principles and Process of Disaster Preparedness and Planning
- Principles and Process of Hazards Mitigation
- Public Administration, Policy and Emergency Management
- Research and Analysis Methods in Emergency Management
- Social Dimensions of Disaster
- Sociology of Disaster
- Special Populations and Vulnerability Issues in Emergency Management
- Technology and Emergency Management
- Terrorism and Emergency Management

centers. Such centers play an important role in knowledge transfer by acting as repositories for hazard-related data, contributing to the growth of research-based knowledge, and serving as contact points for emergency managers seeking ways to upgrade their knowledge and skills. Better-established disaster-related centers, such as the Disaster Research Center (founded in 1963 at Ohio State University and now located at the University of Delaware) and the Natural Hazards Research and Applications Information Center (established in 1976 at the University of Colorado at Boulder), have been joined by over a score of other research entities around the country. Table 6.2 provides a listing of U.S. centers whose work focuses on the social-scientific aspects of hazards, disasters, and emergency management. Some of these research units, such as the Hazard Reduction and Recovery Center at Texas A&M University, conduct studies over a range of different hazards while others, such as the International Hurricane Center at Florida International University, tend to concentrate on the study of particular types of hazards. Some centers, such as the earthquake research consortia centered at SUNY Buffalo, at Illinois, and at Berkeley, include social science research as one element in a larger program; others concentrate exclusively on social-scientific topics.

Other developments have also contributed to the recognition of emergency management as a distinctive field of expertise, as well as to the sharing of information among researchers and practitioners. Interest on the part of organizations such as the International City/County Management Association helped disaster issues attain visibility among local government executives and administrators. Meetings such as the annual National Hurricane Conference and the conference sponsored by the Association of State Floodplain Managers provide venues in which practitioners and researchers can focus on the problems associated with particular hazards. The Natural Hazards Workshop at the University of Colorado, which has been held annually since 1976 and has grown in size each year, was originally begun specifically to bridge the gap between academically-based researchers and hazard management practitioners. In keeping with this objective, its organizers discourage highly technical presentations in favor of more informal panel sessions. The workshop has been so successful in part because it intentionally combines the transmission of academic knowledge and practical lessons with social activities and networking opportunities.

Recent years have also seen the growth of specialized journals focusing on hazard management and disaster-related topics, again ranging

TABLE 6.2 College- and University-Based Disaster and Emergency Management Research, Training, and Information Centers

- California State University – Chico: Center for Hazards Research
- Charleston Southern University: Earthquake Education Center
- Clark University: George Perkins Marsh Institute, Center for Technology, Environment, and Development (CENTED)
- Colorado State University: Hazards Assessment Laboratory
- Florida International University: International Hurricane Center
- George Washington University: Institute for Crisis, Disaster, and Risk Management
- Millersville University: Social Research Group
- New York Medical College: Center for Psychological Response in Disaster Emergencies
- Southwest Texas State University: The James and Marilyn Lovell Center for Environmental Geography and Hazards Research
- State University of New York at Buffalo: Multidisciplinary Center for Earthquake Engineering Research (MCEER)
- Texas A&M University: Hazard Reduction and Recovery Center (HRRC)
- University of Arkansas – Little Rock: Arkansas Center for Earthquake Education and Technology Transfer
- University of California – Berkeley: Continuing Education in Business and Management – Courses and Certification for Emergency Preparedness Managers
- University of California – Berkeley: Pacific Earthquake Engineering Research Center (PEER) and National Information Service for Earthquake Engineering (NISEE)
- University of California – Los Angeles: Center for Public Health and Disaster Relief
- University of California – Riverside: Emergency Management Programs
- University of Colorado – Boulder: Natural Hazards Research and Applications Information Center (NHRAIC)
- University of Delaware: Disaster Research Center
- University of Illinois at Urbana-Champaign: Mid-America Earthquake Center (MAE)
- University of Louisville: Center for Hazards Research and Policy Development
- University of Maryland – Baltimore County: Emergency Health Services Department
- University of Memphis: Center for Earthquake Research and Information (CERI)
- University of New Mexico: Health Sciences Center School of Medicine, Center for Disaster Medicine
- University of New Orleans: Environmental Social Science Research Institute
- University of North Carolina – Chapel Hill: Center for Urban and Regional Studies
- University of North Texas: Emergency Administration and Planning Institute
- University of Pennsylvania: Wharton Risk Management and Decision Processes Center
- University of South Carolina: Hazards Research Laboratory (HRL)

Source: Natural Hazards Research and Applications Information Center

from the more academic and research-based to the more applied and user-oriented. These journals include the *International Journal of Mass Emergencies and Disasters*, published by the International Sociological Association's Research Committee on Disasters; *Disasters*; *Natural Hazards*; the *Journal of Contingencies and Crisis Management*; *Natural*

Hazards Review; *Organization and Environment* (formerly the *Industrial and Environmental Crisis Quarterly*); *Risk Analysis*; *Environment*; *Disaster Management*; *Disaster Recovery Journal*; and *Earthquake Spectra*, the official journal of the Earthquake Engineering Research Institute. Researchers and practitioners also have access to a number of newsletters published both by academic institutions (e.g., the Natural Hazards Center's *Natural Hazards Observer*, the most widely-circulated newsletter in the field) and by practice-oriented organizations such as IAEM. The proliferation of specialized Internet web sites has further aided in the diffusion of disaster-related information, both nationally and internationally.

When the field of disaster research began in the early 1950s, the local civil defense director was likely to be a retired military man operating part-time out of a small office that was both physically removed from and programmatically marginal to centers of community decision making. The civil defense office, which at that time spent more time on war-related crisis planning than on disasters, typically lacked both resources and ties to other governmental units. The civil defense office was a place where people generally went to finish out their careers. Disasters were given a low priority by civil defense and other public safety agencies, except on those occasions when disasters actually did strike.

Although disaster readiness cannot be said to have soared to the top of the political agenda in communities around the country, the field of emergency management certainly enjoys greater prestige today than ever before. At the federal level, the director of FEMA was accorded cabinet rank during the Clinton administration. At the local level, it is now common for the emergency management director to report directly to the local chief executive. Instead of having to make do in a small office in the basement of the fire department, as might once have been the case, today's emergency manager typically has much greater visibility and more resources at his or her disposal. As a consequence of these kinds of changes—the trend toward greater professionalization discussed above and the existence of a career ladder in the emergency management profession—the field is now poised to attract more well-trained, motivated, and ambitious individuals than ever before.

CONCLUDING OBSERVATIONS

Disaster planning and response activities do not take place in a vacuum, but rather are shaped by broader institutional and societal forces. In this chapter, we have discussed a number of those forces, including social, economic, and cultural factors, new developments in technology, and the major shift that has occurred towards greater professionalization of the emergency management field. The impact of these contextual factors and trends has yet to be studied systematically, and research is needed in order to better understand how they affect loss-reduction practices and emergency management effectiveness.

Since the time of the first assessment, basic research in the geologic and atmospheric sciences has led to a better understanding of the physical processes associated with hazards, leading in turn to improvements in vulnerability analysis. Information campaigns have been launched to improve public and organizational preparedness. Better methods of detecting disaster events as they develop—analyzing data, identifying areas and populations that are at risk, and communicating warning information—offer promise for reducing disaster-related mortality and morbidity. It seems reasonable to assume that organizational learning, sound planning principles, education, professionalization, and improved decision-support technologies have enhanced emergency management capacity in many communities around the country, although more research is needed to determine what works, when it works, and, equally important, why it works.

These changes should mean that U.S. society is better able than ever to cope with the threat and the occurrence of disasters. However, while progress has been made, particularly in the areas of public awareness and public sector response effectiveness, anticipating disaster-related problems and responding effectively still present major challenges, and the losses associated with disasters continue to escalate alarmingly. Besides indicating that there is still more to learn and much room for improvement with respect to both preparedness and response, recent disaster experiences also suggest that there is something fundamentally flawed in our society's overall approach to loss reduction. We turn to this in Chapter Seven.

Where Do We Go From Here? Improving What We Know About Disasters While Coping with Them More Effectively

I N THE PRECEDING CHAPTERS, we have assessed research advances in the areas of disaster preparedness and response and have identified social, economic, and other factors that influence those activities. Focusing on social units ranging from the individual and household through other levels of analysis, including organizations, social networks, communities, and the supra-community level, the assessment considered what we now know and what we still do not know about preparing for and responding to disasters. Included were discussions on commonly-used explanatory models, research methods and methodological issues, and on substantive topics ranging from household evacuation and sheltering to federal disaster management policy. This overview has shown that the period since the first assessment of research on natural hazards has been marked by substantial growth in the research literature and corresponding improvements in scientific understanding of emergency preparedness and response activities. More sophisticated methodological approaches have been used, key social processes such as warning response and post-impact mobilization have been studied in greater depth,

243

and a wider range of hazard agents and community settings has been examined. With respect to many topics in the areas of preparedness and response, research findings and conclusions now have significantly more solid empirical support than they did in the past.

There also is considerable evidence that research is having a practical impact on the ways in which disasters are managed. Many of the things researchers have learned about preparedness and response behaviors have been incorporated into crisis management training programs and into courses in institutions of higher education, so that they are now taken into account by a new generation of better-informed emergency personnel. For example, the notion that panic is generally not a problem in disaster situations and that emergency managers need not concern themselves with how to avert it is taught in courses on emergency management and routinely appears in publications that are read by practitioners. Risk communications research has helped produce better warning systems. Agencies responsible for issuing hazard advisories take social scientists' recommendations into account and occasionally involve them directly in developing those warning messages. Emergency management personnel not only know what terms such as emergence and convergence mean; they expect them to occur in disaster situations and have devised ways of dealing with them.

All of these developments are signs of real progress in the growth and transfer of knowledge. Nevertheless, our review has also identified major deficiencies in the knowledge base and occasionally has led to some rather disheartening conclusions. In some cases, the problem is that researchers think they know more than they actually do, because they have accepted conclusions from earlier research without closer examination. Specifically, we do not really know the extent to which social and behavioral patterns identified in some types of disasters generalize to others, because those factors have not been studied to a sufficient degree. In other cases, major questions exist that have still not yet been systematically addressed. Finally, there are areas about which we know a considerable amount but that still require further elaboration.

The literature also lacks balance in its coverage across different hazards and units of analysis. Moving from the household and group through organizational, community, regional, and national levels, the empirical research base becomes scantier as the unit of analysis becomes broader. For example, 23 published studies have revealed quite a bit about household preparedness for earthquakes (Lindell and Perry, 2000), but much less is known about what households do to prepare for other types of disasters or about organizational, interorganizational, and community preparedness generally.

More emphasis needs to be placed on exploring in depth issues that researchers assume—perhaps incorrectly—have already been settled. At the organizational and community levels, for example, can we really show that pre-disaster preparedness makes for a more effective and coordinated response when disaster strikes? If so, in what ways does preparedness matter? To what extent does pre-event preparedness actually contain disaster losses? Is preparedness cost-effective? Researchers and practitioners think so, and they say so, but evidence for such claims is piecemeal and indirect. Large-scale studies are needed to systematically examine the impact of emergency preparedness on the effectiveness of emergency response activities while controlling for differences in disaster impacts and community characteristics. Similarly, researchers have moved toward seeing natural and technological disasters as quite different in the effects they have on individuals and communities (see, for example Freudenburg, 1997, who makes that argument) but, as we observed in Chapter Six, there are many alternative ways of thinking about the natural/technological hazards distinction. The larger point is that the disaster research field continues to take many "truths" for granted that actually have yet to be empirically established.

Additionally, our understanding of preparedness and response would improve if research could move beyond concentrating only on the United States and a handful of other Western countries. What is known about disaster preparedness and response in other societies is rudimentary to say the least, and true cross-national comparative research is exceedingly rare. There are many practical lessons to be learned through in-depth studies of how other societies manage their vulnerability to environmental hazards. Such a focus is particularly necessary as researchers seek to identify sustainable development strategies and ways of making families and communities more resilient during disaster situations. Conducting more comparative research would also provide a needed impetus toward theoretical advancement.

The inability to place the behavioral phenomena associated with emergency preparedness and response into a broader context constitutes another major deficiency in the research literature. We have argued that research must advance beyond considering disasters as isolated, unusual events and instead investigate the ways in which disasters and their management are formed by the social order itself. As we have shown throughout this book, broader features of United States society have helped shape the manner in which the society attempts to cope with disasters. Patterns of disaster-related social and organizational behavior cannot be understood fully without considering the cultural assumptions in which they are rooted, the fragmented and shifting institutional framework within which they have evolved, and the social and economic context within which they are undertaken. As is the case with other societal problems, ways of framing hazards conceptually and dealing with them practically are conditioned both by enduring social and cultural patterns and by social change. Our discussions have shown that, in analyzing how social units deal with the challenges posed by environmental hazards, we must not look narrowly at the disasters themselves but rather must take into account a variety of other factors. These factors include pervasive social inequities that often block access to both the knowledge and the resources people need to protect themselves and avoid disaster losses and an intergovernmental system that, while extraordinarily rich in information, expertise, and monetary resources, seems incapable of acting in a coherent fashion to manage hazards. We also must consider political actors and economic interests that until recently have not even paid lip service to the idea that more can be done to manage hazards. We need to investigate why it is that the same society that acts generously to aid victims when disasters strike also continues to engage in practices that cause losses to mount, supports the right of individuals to put their lives and property in harm's way, and complains about government's interference in private decision making, even when those decisions ultimately contribute to escalating disaster losses. Research is also needed to explore and understand the societal processes that *produce* risk because, without that kind of knowledge, it will be virtually impossible to reduce it.

Along these same lines, we need to take a closer look at the ways in which social diversity and inequality affect patterns of disaster preparedness and response. U.S. society is characterized by increasing cultural and language diversity, growth in size of the elderly and disabled populations, and steady increases in income disparity. All of these factors serve to increase the number of people who lack access to hazard information and to the resources they need to protect themselves against disasters. Recently the U.S. has begun to restrict access to many social programs, including disaster assistance, to both undocumented and many legal immigrants. What implications will these kinds of changes have for overall preparedness levels and response capacity? Will efforts to step up preparedness and make households and communities more resilient make headway in the face of the other pressing problems many households and communities face?

A key theme running through this book is that the past two decades have seen a significant shift in the ways in which researchers characterize and explain disasters. Many of these changes are connected to larger trends in social science theory and research. In the sections that follow, we discuss several thematic areas in which disaster research both draws upon and informs theoretical developments in the social sciences.

DISASTER RESEARCH AND SOCIAL THEORY

Theorizing in the disaster field has mirrored broader theoretical transitions in the social sciences in a number of ways. First, functionalist and systems-oriented perspectives on disaster-related phenomena now co-exist with newer theoretical approaches. The field of disaster research began 50 years ago by conceptualizing disasters as external forces that impinge on cohesive social systems. The hazards paradigm put forth in the first assessment expanded that view, emphasizing the idea that disasters originate as much from human actions as from forces in the physical environment. More recent theory and research have further emphasized how social inequality and diversity affect disaster vulnerability. The image of a harmonious social system has been augmented by a view of preparedness and response behavior that emphasizes socially-structured differences in power, differential access to resources, and patterns of competition and conflict that are only temporarily suspended when disaster strikes. This change is also evident in the related trend away from consensus-oriented models of society and social behavior and the new emphasis on conflict, competition, and social inequality as factors to consider in hazards research.

Again following more general trends in the social sciences, recognition has grown that gender and ethnicity, along with social class, are major stratifying forces in society whose influence also operates in disaster situations. There is now a much greater awareness within the research community that hazard vulnerability is accompanied by an inability to prepare and to respond effectively when disaster strikes, and that these patterns are in turn related to broader patterns of social and economic inequality. This tends to be the case whether the focus is on households, communities, or entire societies. Societies characterized by extreme differences between rich and poor will see those inequities played out in variations in coping capacity and in disaster-induced losses. The reason for this is clear: position within the stratification system affects access to preparedness and response resources and influences the ability to recover from disaster victimization. People for whom everyday life is an ongoing crisis are not likely to be able to protect themselves against the intermittent crises that disasters produce, even if they would like to be able to do so. Intersocietal differences in wealth and power also are reflected in variations in disaster vulnerability, with less-well-off countries suffering disproportionately when disasters strike. These same countries lack the institutional capacity to protect their populations and to respond effectively when they experience a disaster, because vulnerability to disaster impacts and the inability to respond effectively are both reflections of the same underlying social disparities (Blaikie et al., 1994; Bolin, 1998).

Reflecting broader theoretical trends, there has also been a shift from essentialist and realist theorizing about hazards to analyses that recognize the processes through which meanings—including the meanings associated with hazards and related "risk objects" (Hilgartner, 1992)—are collectively generated and assigned. While no disaster researcher is misguided enough to argue that natural and technological hazards are mere illusions, the literature has shown a growing appreciation for the differing ways in which individuals, groups, and societies define danger and harm. Some definitions, such as those involving well-known and common natural hazards, are more institutionalized in the culture and social order and thus less problematic to manage. Others, such as hazards and impacts associated with chronic toxic hazards, are more ambiguous and contested. Similarly, the putative causes of disaster victimization are themselves social constructions, and the boundaries between what is "natural" and what is the product of human action are subject to continual revision.

The field of disaster research also has anticipated changes that have later been incorporated into social theory. For example, for nearly a generation scholars have criticized the manner in which social science theorizing has overemphasized the structured aspects of social behavior while downplaying the role of human agency (see, for example, Giddens, 1984). Disaster research has been criticized on the same grounds for conceptualizing behavior in disaster situations as scripted by roles and driven by norms (Bolin, 1998). At the same time, perhaps because disasters themselves are highly fluid social occasions that breach existing structures, those who study disasters have also been better able than researchers in many other social science fields to recognize emergence, improvisation, and other manifestations of agency in organized social action. This long research tradition includes an extensive literature on emergent groups and on the structuring of organized action in emergencies, as well as newer work on the ways in which members of marginalized populations have mobilized to demand services, fair treatment, and recognition of their concerns in the aftermath of disasters. Indeed, as Robert Bolin has shown in his research on the 1994 Northridge earthquake (1998), the occurrence of a disaster can expand opportunities for community-based organizations and create new avenues for change. One of the key contributions of research on disaster response has been its ability to document the ways in which pre-structured, procedurally-defined, routinized activity intermingles with innovative, collectively-devised behaviors in the disaster context. Indeed, research on disasters may well provide the most compelling example of how all social life consists of a merging of agency with structure.

This appreciation of the role of emergence and improvisation stems in part from the close relationship that has existed between disaster research and the field of collective behavior (Wenger, 1987). Many scholars who conduct research on post-disaster response behaviors are also interested in other types of collective behavior phenomena, and considerable cross-fertilization has occurred between the two areas of specialization (Dynes and Tierney, 1994). The use of emergent norm theory (Turner and Killian, 1987) in explanations of warning response behaviors, which were discussed in Chapter Three, is one example, as is the concern with the conditions that give rise to panic. Many of the behaviors that commonly develop under conditions of disaster threat and impact—including increased information-seeking and the transmission of rumors, evacuation behavior, the improvised activities of groups that form to carry out emergency tasks, and the convergence of people to disaster-stricken areas—fall within the domains of both disaster research and collective behavior scholarship. Response to pseudo-disaster threats and false warnings such as the 1990 Iben Browning earthquake "prediction" clearly lend themselves to analyses using collective behavior frameworks (Tierney, 1994). The notion that collective behavior involves emergence both of norms and of new forms of social organization is grounded in research on behavior in disaster situations (Weller and Quarantelli, 1973; Stallings and Quarantelli, 1985). Textbooks on collective behavior now routinely include sections on disaster behavior and disaster research (see, for example, Goode, 1992, and Marx and McAdam, 1994), again indicating the affinity that exists between the two research specialties.

Recent years have also seen theoretical convergence between disaster studies and more general research on the environment. The first assessment established a linkage between research on disasters and the broader field of environmental studies, and that linkage has been strengthened. Environmental researchers have increasingly turned their attention to study of natural and technological disasters, as well as to research on chronic exposure to toxic hazards (see, for example, Gramling and Freudenburg, 1992, and Hannigan, 1995). Along with environmental degradation and resource depletion, disaster impacts are now commonly seen as part of a complex of negative environmental outcomes resulting from policies that emphasize growth at the expense of safety and from the operation of political-economic forces that depend on the exploitation of natural and environmental resources.

PREPAREDNESS AND RESPONSE IN CONTEXT: SEEING THE LARGER PICTURE

Perhaps the strongest developing theme in the disaster literature, also linked with research on the environment, is one that connects disasters with the broader concept of sustainability. This broader framework for theorizing about disasters, which has its roots in many of the ideas developed as part of the first assessment and which is articulated in various ways by contributors to the second assessment, argues that the same economic and social processes that are implicated in unsustainable patterns of development and in the depletion of natural resources also give rise to more frequent disaster events and escalating losses. As Beatley has noted in *Cooperating With Nature* (1998), researchers continue to document the numerous ways in which development patterns that are characteristic of U.S. society (and, we would add, those of the present-day world system) virtually ensure that the impacts of normal environmental fluctuations will become increasingly disastrous, generating ever higher losses and more severe social disruption. The influential volume *At Risk* (Blaikie et al., 1994) makes the same general point: disasters represent the convergence of unsustainable development practices, vulnerable populations, and—finally—some event in the physical environment that acts as a trigger, causing damage, casualties, and losses.

Scholarship in the field of environmental studies provides insights into the social forces that drive unsustainable development practices. A prominent theme in that literature is that environmental destruction results from both a national and international "treadmill of production" that is driven by two processes: increasing reliance on technology to provide economic outputs and the dominance of interests that promote economic growth, regardless of the fact that ecosystems will be harmed (Schnaiberg and Gould, 1994). Operating unchecked, the treadmill treats the natural environment merely as something to be used in the productive process, consuming non-renewable resources at an ever-accelerating pace, and then dumping the by-products of production back into nature in the form of toxins and other wastes. Although there are countervailing pressures, such as those arising from environmental and limited-growth movements, the promotion of economic growth remains an overarching priority at the local, state, national, and international levels. The costs and negative effects of growth, which include pollution and other environmental problems, are borne by marginalized populations (e.g., exploited and displaced workers, people living near toxic sites) and by succeeding generations. Ever-increasing disaster losses are part of this legacy.

Many of the steps that can be taken to avoid exposure to hazards and to prevent disaster damage are the responsibility of local decision makers, yet development pressures are invariably most intense at the local level. Environmental scholars (see for example, Cable and Cable, 1995; Buttel, 1997) note that the notion of a treadmill of unsustainable economic activity is conceptually linked to the concept of local "growth machines" and "growth coalitions" (Logan and Molotch, 1987). As described by Buttel, growth coalitions consist of commercial, real estate, and other related interest groups such as tourism boosters whose activities are "focused on the expectation that each will directly or indirectly benefit from growth in public subsidies to and private investments in infrastructure, civic capital, construction, and related activities that help to attract people, employers and jobs to a local area" (1997: 47). The goal of such coalitions is increasingly intensive land use and development, regardless of whether that development takes place in hazardous areas. Controlling development in high-hazard locations is invariably difficult because of the immense political power wielded by pro-growth interests. Such patterns persist even following major disasters. As May and Deyle have noted (1998: 62), various studies on post-disaster recovery have documented "the reluctance of local governments to significantly restrict land use in hazardous areas even when the risks of such land use have been vividly demonstrated."

Seen in this context, disasters are part of a continuum of negative environmental impacts that result from unsustainable development practices. The effects of hazard agents are so pronounced because human settlements are based upon principles of short-term growth and profits for privileged segments of the population instead of safety and sustainability for the society as a whole.

Another way unsustainable development helps to produce disastrous consequences is by compromising the ability of the natural environment to contain the effects of triggering events. For example, as Beatley (1998) has argued, the extensive networks of roadways and other paved surfaces and dense concentrations of buildings that characterize today's built environment undermine the land's capacity to absorb flooding. Moreover, the walls, levees, and other public works that make up modern flood control systems only set the stage for larger future flood losses. The 1993 Midwest floods are one recent example of the consequences of this approach to managing flood hazards. Writing on the history and impacts of flood control in the Midwest region, Wilkins noted (1996: 220):

> The first river control study in the 1850s provided the template for much of what was to come in the ensuing 150 years. The initial study of 1849 concluded that the river should be controlled with a 'levees-only' policy, a refuge in technology that was supplemented after the massive 1927 flood . . . by other structural measures such as reservoirs, fuse-plug levees, floodways and channel improvements. Well before the flood of 1993, this reliance on technology and physical structures had resulted in an extensive network of federally constructed levees augmented by thousands of agricultural levees built to much less exacting standards, a riverbed that had been dredged and channeled for many years, and, within the confines of St. Louis itself, a river wall 49 feet above normal river flood levels through the downtown corridor that would come within a few inches of being topped by the overflowing Mississippi during the flood of 1993. While there is still enormous technical dispute on precisely how much impact this reliance on technological control had on mitigating and exacerbating the consequences of 'a lot of rain,' there is little dispute that Missouri provides a microcosm through which to study the convergence of historic, economic, and social conditions in conjunction with a natural disaster of historic proportions.

This structurally- and technologically-focused strategy for managing the flood threat is itself a consequence of economic, political, and institutional forces that promote growth by directly or indirectly subsidizing development. It has long been pointed out that rather than providing long-term protection, the "engineered structural works" approach can make more-catastrophic future losses more likely by encouraging development in unsafe areas. Investors reap the benefits of development while the hidden costs in the form of disaster losses are deferred, to be paid later by disaster victims and taxpayers. While the question of whether other strategies for controlling Mississippi River flooding would have reduced losses resulting from the 1993 floods still has not been settled, the overall consensus among researchers is that the habitual overreliance on large-scale public works as a means to control flooding is a key factor in escalating flood losses. Citing an earlier government report, for example, Burby observed (1998: 8) that "fully two-thirds of national losses in flooding result from catastrophic events that exceed the design limitations of engineering works that are relied on to provide safety."

Viewing the threat of disasters from the perspective of sustainability, the key to protecting society against future disaster losses lies in reversing current short-sighted development practices and substituting alternative approaches that are sustainable in the longer term. As Geis has noted, a close linkage exists between community development and planning policies and disaster vulnerability. The seeds of future disasters lie in "community development patterns, transportation and utility design and configuration, relationship between the built and the natural environments, patterns of open space, housing and neighborhood design, and building group configuration and location" (Geis, 1996: 3). Superimposed upon community residents' social vulnerabilities, these physical vulnerabilities are major factors in producing disaster losses. It follows, then, that in order to lessen the consequences of future events, communities must develop in ways that are sustainable while simultaneously addressing issues of social vulnerability.

In *Cooperating With Nature* (1998) Beatley argued that sustainable and disaster-resistant communities are those that simultaneously pursue both safety and other civic goals using a diverse set of strategies. First, they minimize the exposure of people and property to natural disasters, recognize ecological limits, and direct their efforts toward enhancing the integrity of ecosystems. Second, they try to promote a deeper understanding of the natural environment and reduce the demands people place on land and resources. Third, they link environmental, social, and economic goals and focus on protecting the community's "ecological capital." Fourth, they replace disjointed, contradictory policies with more comprehensive ones that seek to address broad community needs, including the need for housing, protection of the environment, and disaster resistance—in a coherent way rather than in isolation. Finally, they view environmental resource conservation and protection against natural hazards in moral and ethical terms, seeking social equity through environmental and hazard policy. Communities are not accustomed to thinking in these ways about hazards, but successful implementation of strategies like these would almost certainly result in a steady if slow decline in disaster losses. What remains to be seen, however, is whether sustainable hazard reduction, integrated into a broader program of sustainable development, can make headway against the powerful societal forces that support current policies and practices.

PREPAREDNESS, RESPONSE, AND SUSTAINABILITY

As a culture, we seem more focused than ever before on disasters. The daily news contains a steady stream of stories about how communities across the country are coping with the latest flood, tornado, wild fire, or chemical release. These accounts invariably focus on the steps taken by official agencies to manage disaster impacts and on stories of individual heroism and courage. Missing from these disaster narratives, which draw upon common cultural themes and media reporting conventions, are discussions of the forces that contribute to the proliferation of crisis events and of what can be done in both the short- and long-term to reduce their frequency and severity. Disasters are portrayed both as societal abnormalities and as discrete events, without reference to the larger societal context. The overall message is that, since disasters are unfortunate if inevitable acts of nature, perhaps the best we can do is cope with them, clean up, provide relief, and go on. Our society has a short attention span. When the emergency period ends, so does the public's interest—until the cycle resumes with the next disaster.

In a related vein, Maskrey (1994) has highlighted the tendency for both the mass media and responding agencies to conceptualize disaster management according to what he terms the "kitsch paradigm." That term refers to the assumption that disasters are best handled through the massive mobilization of material and human resources by official response agencies. Although Maskrey's observations were based on research in Latin America, they apply equally to the United States. Until very recently both social science research and government policy have focused on disaster preparedness, response, and short-term relief as if those activities constitute the core of what needs to be done to protect the public against natural and technological hazards. U.S. society has followed the practice of "fix upon failure," mobilizing massively when disaster strikes, providing material aid to victims, and then restoring damaged communities as rapidly as possible, even if that meant providing little protection against future damage. The nation's alarmingly expanding expenditures on disaster response and post-disaster aid are one indication of the extent to which this society deals with disasters in a reactive and event-focused, rather than a proactive and comprehensive, fashion.

More broadly, but again in keeping with dominant cultural emphases, science and technology continue to be seen as providing the main solution to managing disasters and their socioeconomic impacts. Thus, we see an emphasis on forecasting hurricanes and floods, using improved technologies to detect the formation of tornadoes, monitoring seismic activity, and (though this goal has proved elusive) predicting earthquakes, as well as, more recently, attempting to harness the power of information technology to address response-related challenges. Indeed, viewers of the recent pre-millennial deluge of television documentaries on disasters might well come away believing that the key to reducing disaster losses lies in obtaining more rapid and accurate scientific information on where and when extreme events will occur rather than in developing societal strategies that protect against loss and disruption. There is little recognition that, besides doing nothing to attack the root cause of disasters, overreliance on technological fixes to disaster-related problems ultimately privileges the entities that control those technologies, typically large governmental agencies and corporations, while excluding those that lack access to technology. Similarly, there is little acknowledgment that gains in science and technology, while important, will have little impact unless they are accompanied by changes in the way society thinks about disasters and the steps that are taken to manage them.

U.S. society's current strategy for dealing with hazards too often parallels its response in other policy arenas. Health and illness are examples that immediately come to mind. The health-care system concentrates on managing acute disease episodes rather than on prevention, relying extensively on heroic forms of intervention and advanced medical technologies. Although it is now increasingly recognized that preventive care, exercise, and sound nutrition make for a healthier society, fighting illness still takes precedence over promoting health. Similarly, society responds to the ever-escalating number of disasters by responding massively and pouring ever-increasing amounts of post-disaster aid into stricken regions, rather than on reducing the need for that aid.

Of course, no reasonable person would argue that resources for responding rapidly in the event of a disaster are unnecessary, or that we should not use the best technologies available for coping with crises. But the response-driven, technological image of hazard management obscures other more viable strategies that would focus on reducing risk rather than on dealing with the consequence of ignoring it. In keeping with the theme of the assessment, we use the term "sustainable hazard and disaster management" to describe these strategies. The overall goals of sustainable hazard and disaster management are to reduce physical, social, and economic vulnerability and to facilitate the effective provision of short-term emergency assistance and longer-term recovery aid. What follows is a brief list of five research-based recommendations for achieving those goals.

Build a consensus that avoiding disasters is preferable to responding to or recovering from them.

While there still needs to be an emphasis on discovering ways to better manage disasters when they occur, an even greater emphasis should be placed on lessening the need for crisis management by reducing the frequency and severity of disaster events. As Donald Geis has argued (1996: 3):

> We can develop and implement the very best emergency management plan possible, the most efficient well planned preparedness plan, respond in the most efficient way possible after a disaster occurs, and execute a sound recovery. But as important and effective as each of these may be, none are nearly as important relative to achieving our primary goal as the process of creating disaster-resistant communities. Neither can any of their functions and roles be optimized in their own right in an emergency management context without this process.

In other words, while enhancing disaster management capabilities, we must also address the root causes of disasters and encourage fundamental change in the hazard adjustment process. People who have learned to wear seat belts, stop smoking, and eat low-fat diets have shown themselves to be capable of changing their behavior in ways that make them safer. What is needed to bring about change in behavior with respect to hazards are broad, society-wide initiatives—similar to the Project Impact initiative that FEMA is currently undertaking but on a much larger scale—to institutionalize mitigation on the political agenda. There will always be a need for effective preparedness, response, and recovery measures, but the overriding goal of emergency management policy and practice should be to reduce the incidence of disasters and thereby decrease the need for managing them. Programs are needed to help people understand why disasters happen, to provide information on their costs to society in terms of deaths, injuries, damage, and economic losses, and most importantly, to emphasize the message that those losses can be reduced. These programs should seek to place disasters on the list of problems that society has succeeded in addressing through sound policies that encourage positive behavior change, and they should convey the message that disaster losses are no more inevitable than dying at an early age from heart disease. Disaster victimization will decline only when people demand protection against hazards in the same way that they now demand automobile, airline, and food safety.

Of course, placing mitigation on the political agenda and keeping it there are by no means strictly a matter of educating individuals. Parallel initiatives are also needed to provide incentives for influential organizational and institutional actors to incorporate disaster-loss avoidance into their ongoing activities. This will necessarily involve providing both rewards for practices that enhance safety and penalties for risk-producing activities. Society sustains so many disasters because too many actors incur too few costs for allowing disasters to occur. Hazard management policies need to recognize that risks will decline when risky choices stop being profitable.

Approach disaster preparedness and response comprehensively.

As discussions throughout this volume have shown, disaster preparedness and response are often fragmented and compartmentalized. Organizations tend to prepare in isolation from one another or to join together only with similar organizations. When they do manage to plan together, officially-designated emergency organizations still tend to ignore those without disaster responsibilities, and public- and private-sector hazard management efforts often proceed on separate tracks. Nongovernmental and community-based organizations that may offer the best avenue for connecting with community residents typically also lack a voice when issues of disaster preparedness are considered. Despite notable improvements, examples of truly integrated community-wide preparedness and response networks are rare.

Fragmentation also is evident across the different phases of the disaster cycle. Response-oriented organizations such as local emergency management agencies frequently lack ties to the community development and building safety departments that have jurisdiction over measures that can mitigate the effects of hazards, as well as to the organizations that would play a role in recovery decision making should a disaster occur. This compartmentalization blocks the free flow of information among parties responsible for different stages in the disaster management cycle, militating against the kinds of action that are needed to reduce the impacts of disasters. One reason that GIS has been embraced so enthusiastically by the hazards community is that it provides a platform for addressing hazard- and disaster-management problems more holistically. However, while GIS has great potential for aiding hazard- and crisis-management decision making, it cannot substitute for the development of cooperative working relationships and policies that focus on reducing disaster losses.

Integrate hazard management into the activities of grassroots community organizations.

Conceptions about how to carry out effective preparedness and response activities must also become broader and more inclusive. In sustainable hazard disaster management, the emphasis should be on relying on indigenous community strengths rather than on hierarchical, centralized management models and on balancing expert knowledge with local knowledge. As the disaster literature documents, a large share of the resources needed to cope effectively with crises reside not in official crisis-relevant organizations but rather in community-based groups, organizations that operate for purposes other than disasters, and within the public at large. Efforts to increase the salience of hazards and disasters among local neighborhood watch organizations, train community residents to respond in disaster situations, and link volunteer groups with official response agencies are part of a positive trend toward thinking more comprehensively about crisis management.

Employ appropriate strategies for managing disasters.

Efforts to prepare for and respond to disasters must be grounded in an understanding of how people and organizations behave during crises. Planning models are doomed to failure when they are based on the assumption that a situation as complex and rapidly-changing as a major disaster can be centrally controlled by a single decision-making entity. In fact, the trend is in the opposite direction: As disasters become larger and more complex, and as the media and technology make information more widely available, the number of entities that can become involved in emergency response also will grow, and crisis decision making will become increasingly decentralized. The "command and control" approach, which never was appropriate for managing disasters, represents a thoroughly outdated way of thinking about crisis response. Instead, policies and plans should conceptualize disaster response as a loosely-coupled set of activities carried out by a highly diverse set of entities: official crisis-relevant organizations, voluntary groups, community-based organizations, emergent citizen groups, and the public at large. Seen in this light, the disaster-related activities of officially-designated emergency agencies actually constitute only a small segment of a very large spectrum of organized crisis activity.

Disaster scholarship (see, for example, Dynes, 1993) also emphasizes the notion that, rather than being seen as troublesome or as impediments to the smooth management of crisis response systems, community residents should be seen as resources that can enhance response capability if allowed to do so. More generally, it should be recognized that people facing environmental hazards require information on what to do to protect themselves, why they should undertake those actions, and how to obtain and provide help in disaster situations. Taking warning response as an example, just as agencies need not fear creating panic if they warn people of impending harm, they also need to understand that people generally do not heed warnings merely because they have been ordered to do so. Rather, they act when they have weighed available warning information and decided that action is prudent and feasible. This same principle holds true for other forms of self-protective action. People will behave in ways that enhance their safety when they understand that it is in their best interests to do so, when they know what they should do, and when they can afford to act. This principle applies whether that action involves retrofitting a house, purchasing hazard insurance, developing a household disaster plan, or adopting any other self-protective measure.

Similarly, emergency planning should be based on appropriate assumptions about individual and group behavior. Response agencies and service providers should not expect people to change longstanding cultural practices and ways of adapting when faced with disaster. Rather, they should seek to better understand those patterns and develop their programs accordingly.

Tailor preparedness and response efforts to the needs and capabilities of those being served.

Households, organizations, and communities vary markedly both in their hazard vulnerability and in their capacity to mitigate, prepare, respond, and recover from disasters. Recognizing these differences in vulnerability and capacity, all hazard management policies and programs should be adapted to needs of specific groups and community settings rather being uniformly applied to all target audiences and service recipients. Perhaps the best way to address the needs of an increasingly diverse population is to involve community residents more directly in program development and service provision. Where it is impossible to avoid standardization and bureaucratic formality, care should be taken to ensure that social and cultural diversity do not act as barriers to service utilization.

Our understanding of both our physical environment and the ways in which environment and society interact remains incomplete. For the foreseeable future, we will be living with the consequences of having steadily if unintentionally created vulnerable communities. While taking every opportunity to reduce this vulnerability, U.S. society must still be ready to respond when disasters strike, as it inevitably will. However, if as a society we succeed in bringing about fundamental changes in the manner in which hazards are perceived and managed, we can all face the unexpected with greater confidence.

References

Acker, J. 1990. "Hierarchies, jobs, bodies: A theory of gendered organizations." *Gender and Society* 4: 139–158.

Acker, J. 1992. "Gendered institutions: From sex roles to gendered institutions." *Contemporary Sociology* 21: 565–569.

Adams, W. C., S. D. Burns, and P. G. Handwerk. 1994. *Nationwide LEPC Survey*. Washington, DC: Department of Public Administration, George Washington University.

Aguirre, B. E. 1988. "The lack of warnings before the Saragosa tornado." *International Journal of Mass Emergencies and Disasters* 6: 65–74.

Aguirre, B. E., D. Wenger, and G. Vigo. 1998. "A test of the emergent norm theory of collective behavior." *Sociological Forum* 13: 301–320.

Aguirre, B., D. Wenger, T. A. Glass, M. Diaz-Murillo, and G. Vigo. 1995. "The social organization of search and rescue: Evidence from the Guadalajara gasoline explosion." *International Journal of Mass Emergencies and Disasters* 13: 93–106.

Aguirre, B. E., W. A. Anderson, S. Balandran, B. E. Peters, and H. M. White. 1991. *Saragosa, Texas, Tornado May 22, 1987: An Evaluation of the Warning System*. Washington, DC: National Academy Press.

Alesch, D. J. and W. J. Petak. 1986. *The Politics and Economics of Earthquake Hazard Mitigation: Unreinforced Masonry Buildings in Southern California*. Boulder, CO: University of Colorado, Institute of Behavioral Science, Program on Environment and Behavior.

Altheide, D. 1976. *Creating News: How T. V. News Distorts Events*. Thousand Oaks, CA: Sage Publications.

Altheide, D. and R. P. Snow. 1979. *Media Logic*. Thousand Oaks, CA: Sage Publications.

Andersen, M. L. and P. H. Collins (eds.). 1998. *Race, Class, and Gender: An Anthology.* Belmont, CA: Wadsworth.

Anderson, W. A. 1965. *Some Observations on a Disaster Subculture: The Organizational Response of Cincinnati, Ohio to the 1964 Flood.* Newark, DE: University of Delaware, Disaster Research Center. DRC Report Series No. 6.

Anderson, W. A. 1969. *Local Civil Defense in Natural Disaster: From Office to Organization.* Newark, DE: University of Delaware, Disaster Research Center. DRC Report Series No. 7.

Anderson, P. S. 1995. "The biggest mutual aid system on earth: The Internet in emergency management." *NCCEM Bulletin*: 7–9.

Anderton, D. L., A. B. Anderson, P. H. Rossi, J. M. Oakes, M. R. Fraser, E. W. Weber, and E. J. Calabrese. 1994. "Hazardous waste facilities: 'Environmental equity' issues in metropolitan areas." *Evaluation Review* 18: 123–140.

Anderton, D. L., J. M. Oakes, and K. L. Egan. 1997. "Environmental equity in Superfund: Demographics of the discovery and prioritization of abandoned toxic sites." *Evaluation Review* 21: 3–26.

Angel, R. and M. Tienda. 1982. "Determinants of extended household structure: Cultural pattern or economic need?" *American Journal of Sociology* 87: 1360–1383.

Anthony, D. 1994. "Managing the disaster." *Fire Engineering* 147: 22–40.

Aron, J. 1990. "Nuclear emergencies." Pp. 197–218 in W. L. Waugh, Jr. and R. J. Hy (eds.) *Handbook of Emergency Management: Programs and Policies Dealing with Major Hazards and Disasters.* New York: Greenwood Press.

Aronoff, M. and V. Gunter. 1992. "Defining disaster: Local constructions for recovery in the aftermath of chemical contamination." *Social Problems* 39: 345–365.

Atsumi, T., T. Sugiman, H. Mori, and I. Yatsuduka. 1996. "Participant observations on volunteer organizations emerging after the Great Hanshin earthquake: Case of the Nishinomiya Volunteer Network and the Local NGOs Coordinating Team for the Great Hanshin earthquake." Pp. 455–462 in *Proceedings of the International Conference on Water Resources and Environmental Research* (Vol. II). Kyoto, Japan.

Auf der Heide, E. 1989. *Disaster Response: Principles of Preparation and Coordination.* St. Louis, MO: C. V. Mosby.

Baca Zinn, M. and B. T. Dill (eds.). 1994. *Women of Color in U.S. Society.* Philadelphia: Temple University Press.

Baker, E. J. 1991. "Hurricane evacuation behavior." *International Journal of Mass Emergencies and Disasters* 9: 287–310.

Balm, R. 1993. "High winds and headlines: Stereotypes of nature in the Hurricane Andrew reporting of three newspapers." Paper presented at the Graduate Geography Student Conference, University of Kentucky, Lexington, KY.

Banerjee, M. M. and D. F. Gillespie. 1994. "Strategy and organizational disaster preparedness." *Disasters* 18: 344–354.

BAREPP/NCEER. 1992. *Findings and Recommendations: Symposium on Policy Issues in the Provision of Post-Earthquake Shelter and Housing.* Buffalo, NY: State University of New York at Buffalo, Multidisciplinary Center for Earthquake Engineering Research.

Barlow, H. D. 1993. "Safety officer accounts of earthquake preparedness at riverside industrial sites." *International Journal of Mass Emergencies and Disasters* 11: 421–436.

Barton, A. H. 1969. *Communities in Disaster: A Sociological Analysis of Collective Stress Situations.* Garden City, NY: Doubleday and Co.

Bates, F. L. and C. Pelanda. 1994. "An ecological approach to disasters." Pp. 145–159 in R. R. Dynes and K. J. Tierney (eds.) *Disasters, Collective Behavior, and Social Organization.* Newark, DE: University of Delaware Press.

Baum, A., R. Fleming, and L. Davidson, 1983. "Natural hazards and technological catastrophes." *Environment and Behavior* 15: 333–354.

Bauman, D. and J. Sims. 1978. "Flood insurance." *Economic Geography* 54: 189–96.

Beady C. H. and R. C. Bolin. 1986. "The Role of the Black Media in Disaster Reporting to the Black Community." Boulder, CO: Institute of Behavioral Science, University of Colorado. Working Paper No. 56.

Beatley, T. 1998. "The vision of sustainable communities." Pp. 233–262 in R. J. Burby (ed.) *Cooperating With Nature: Confronting Natural Hazards with Land-Use Planning for Sustainable Communities.* Washington, DC: Joseph Henry Press.

Beck, U. 1992. *Risk Society: On the Way to an Alternative Modernity.* Thousand Oaks, CA: Sage.

Beck, U. 1995a. *Ecological Enlightenment: Essays on the Politics of the Risk Society.* Atlantic Highlands, NJ: Humanities Press.

Beck, U. 1995b. *Ecological Politics in an Age of Risk.* Cambridge: Polity Press.

Been, V. and F. Gupta. 1997. "Coming to the nuisance or going to the barrios? A longitudinal analysis of environmental justice claims." *Ecology Law Quarterly,* February: 1–56.

Beggs, J., V. Haines, and J. Hurlbert. 1996. "The effects of personal network and local community contacts on the receipt of formal aid during disaster recovery." *International Journal of Mass Emergencies and Disasters* 14: 57–78.

Berke, P. R. 1991. "Risk, politics, and vertical shelter policy." Pp. 123–167 in Ruch, C., H. C. Miller, M. Haflich, N. M. Farber, P. R. Berke, and N. Stubbs (eds.) *The Feasibility of Vertical Evacuation.* Boulder, CO: University of Colorado, Institute of Behavioral Science.

Berke, P. R., T. Beatley, and S. Wilhite. 1989. "Influences on local adoption of planning measures for earthquake hazard mitigation." *International Journal of Mass Emergencies and Disasters* 7: 33–56.

Blaikie, P. T. Cannon, I. Davis, and B. Wisner. 1994. *At Risk: Natural Hazards, People's Vulnerability, and Disasters.* London: Routledge.

Blau, J. and P. Blau. 1982. "The cost of inequality: Metropolitan structure and violent crime." *American Sociological Review* 47: 114–129.

Blocker, T. J. and D. E. Sherkat. 1992. "In the eyes of the beholder: Technological and naturalistic interpretations of a disaster." *Industrial Crisis Quarterly* 6: 153–166.

Blocker, T. J., E. B. Rochford Jr., and D. E. Sherkat. 1991. "Political responses to natural hazards: Social movement participation following a flood disaster." *International Journal of Mass Emergencies and Disasters* 9: 367–382.

Bogard, W. C. 1988. "Bringing social theory into hazards research: Conditions and consequences of the mitigation of environmental hazards." *Sociological Perspectives* 31: 147–168.

Boileau, A., B. Catarinussi, B. Della Zotti, C. Pelanda, R. Strassoldo, and B. Tellia. 1979. *Friuli: La Provca del Terremota.* Milan: Franco Angeli.

Boje, D. M. and D. A.Whetten. 1981."Effects of organizational strategies and constraints on centrality and attributions of influence in interorganizational networks." *Administrative Science Quarterly* 26: 378–395.

Bolin, R. C. 1982. *Long-Term Family Recovery from Disaster.* Boulder, CO: University of Colorado, Institute of Behavioral Science, Program on Environment and Behavior.

Bolin, R. C. 1993. *Household and Community Recovery After Earthquakes.* Boulder, CO: University of Colorado, Institute of Behavioral Science, Program on Environment and Behavior.

Bolin, R. C. 1994. "Post-disaster sheltering and housing: Social processes in response and recovery." Pp. 115–127 in R. R. Dynes and K. J. Tierney (eds.) *Disasters, Collective Behavior, and Social Organization.* Newark, DE: University of Delaware Press.

Bolin, R. C., with L. Stanford. 1998. *The Northridge Earthquake: Vulnerability and Disaster.* London: Routledge.

Bolin, R. C. and P. Bolton. 1986. *Race, Religion, and Ethnicity in Disaster Recovery.* Boulder, CO: University of Colorado, Institute of Behavioral Science, Program on Environment and Behavior.

Bolin, R. C. and D. J. Klenow. 1988. "Older people in disaster: A comparison of black and white victims." *Journal of Aging* 26: 29–45.

Bolin, R. C. and L. Stanford. 1990. "Shelter and housing issues in Santa Cruz County." In R. Bolin (ed.) *The Loma Prieta Earthquake: Studies of Short-Term Impacts.* Boulder, CO: University of Colorado, Institute of Behavioral Science, Program on Environment and Behavior.

Bolin, R. C. and L. Stanford. 1993. "Emergency sheltering and housing of earthquake victims: The case of Santa Cruz County." Pp. B43–B50 in P. A. Bolton (ed.) *The Loma Prieta, California, Earthquake of October 17, 1989—Public Response.* U.S. Geological Survey Professional Paper 1553–B. Washington, DC: U.S. Government Printing Office.

Bolin, R. C. and L. Stanford. 1998. "The Northridge earthquake: Community-based approaches to unmet recovery needs." *Disasters* 22: 21–38.

Bolin, R., M. Jackson, and A. Crist. 1998. "Gender inequality, vulnerability, and disaster: Issues in theory and research." Pp. 27–44 in E. Enarson and B. H. Morrow (eds.) *The Gendered Terrain of Disaster.* Westport, CT: Praeger.

Bolton, P., E. B. Liebow, and J. L. Olson. 1992. "Community context and uncertainty following a damaging earthquake: Low-income Latinos in Los Angeles." Paper presented at the annual meeting of the Society for Risk Analysis.

Bosworth, S. L. and G. A. Kreps. 1986. "Structure as process: Organization and role." *American Sociological Review* 51: 699–716.

Bourque, L., L. A. Russell, and J. D. Goltz. 1993. "Human behavior during and immediately after the earthquake." Pp. 3–22 in Patricia Bolton (ed.) *The Loma Prieta, California, Earthquake of October 17, 1989—Public Response.* U.S. Geological Survey Professional Paper 1553–B. Washington, DC: U.S. Government Printing Office.

Botterel, A. 1995–96. "Network technology in the practice of emergency management." *Australian Journal of Emergency Management* 10: 43.

Brickman, R., S. Jasanoff, and T. Ilgen. 1985. *Controlling Chemicals: The Politics of Regulation in Europe and the U.S.* Ithaca, NY: Cornell University Press.

Britton, N. R., C. C. Moran, and B. Correy. 1994. "Stress coping and emergency disaster volunteers: A discussion of some relevant factors." Pp. 128–144 in R. R. Dynes and K. J. Tierney (eds.) *Disasters, Collective Behavior, and Social Organization.* Newark, DE: University of Delaware Press.

Brouillette, J. R. and E. L. Quarantelli. 1971. "Types of patterned variation in bureaucratic adaptations to organizational stress." *Sociological Quarterly* 41: 39–46.

Brunacini, A. 1985. *Fire Command.* Quincy, MA: National Fire Protection Association.

Bullard, R. D. 1990. "Ecological inequities and the New South: Black communities under siege." *Journal of Ethnic Studies* 17: 101–115.

Bullard, R. D. 1994. *Dumping in Dixie: Race, Class, and Environmental Quality.* Boulder, CO: Westview Press.

Burby, R. (ed.) 1998. *Cooperating With Nature: Confronting Natural Hazards with Land-Use Planning for Sustainable Communities.* Washington, DC: Joseph Henry Press.

Burby, R. J., B. A. Cigler, S. P. French, E. J. Kaiser, J. Kartez, D. Roenigk, D. Wiest, and D. Whittington. 1991. *Sharing Environmental Risks: How to Control Governments' Losses in Natural Disasters.* Boulder, CO: Westview Press.

Burton, I., R. W. Kates, and G. F. White. 1978. *The Environment as Hazard.* New York: Oxford University Press.

Buttel, F. 1976. "Social science and the environment: Competing theories." *Social Science Quarterly* 57: 307–323.

Buttel, F. 1997. "Social institutions and environmental change." Pp. 40–54 in M. Redclift and G. Woodgate (eds.) *The International Handbook of Environmental Sociology.* Cheltenham, UK: Edward Elgar.

Cable, S. and C. Cable. 1995. *Environmental Problems, Grassroots Solutions: The Politics of Grassroots Environmental Conflict.* New York: St. Martin's Press.

Canter, D. 1980. *Fires and Human Behaviour.* London: John Wiley and Sons.

Capek, S. 1993. "The 'environmental justice' frame: A conceptual discussion and an application." *Social Problems* 40: 5–24.

Caplow, T., H. M. Bahr, and B. A. Chadwick. 1984. *Analysis of the Readiness of Local Communities for Integrated Emergency Management Planning.* Charlottesville, CA: United Research Services, Inc.

Carter, T. M. 1980. "Community warning systems: The relationships among the broadcast media, emergency service agencies, and the National Weather Service." Pp. 214–228 in *Disasters and the Mass Media: Proceedings of the Committee on Disasters and the Mass Media Workshop.* Washington, DC: National Academy of Sciences/National Research Council.

Carter, N. 1991. *Disaster Management: A Disaster Manager's Handbook.* Manila: Asian Development Bank.

Carter, M., J. Clark, R. Leik, and G. Fine. 1977. "Social factors affecting dissemination of and response to warnings." Paper presented at the Eleventh Technical Conference on Hurricanes and Tropic Meteorology, Miami Beach, FL.

Chetkovich, C. A. 1997. *Real Heat: Gender and Race in the Urban Fire Service.* New Brunswick, NJ: Rutgers University Press.

Clarke, L. B. 1985. "The origins of nuclear power: A case of institutional conflict." *Social Problems* 32: 473–487.

Clarke, L. B. 1989. *Acceptable Risk? Making Decisions in a Toxic Environment.* Berkeley: University of California Press.

Clarke, L. B. 1990. "Oil spill fantasies." *Atlantic Monthly,* November: 65–77.

Clarke, L. B. 1993. "The disqualification heuristic: When do organizations misperceive risk?" *Research in Social Problems and Public Policy* 5: 289–312.

Clarke, L. B. 1999. *Mission Improbable: Using Fantasy Documents to Tame Disaster.* Chicago: University of Chicago Press.

Clarke, L. B. and J. F. Short. 1993. "Social organization and risk: Some current controversies." *Annual Review of Sociology* 19: 375–99.

Clary, B. B. 1985. "The evolution and structure of natural hazard policies." *Public Administration Review* 45: 20–28.

Clausen, L., P. Conlon, W. Jager, and S. Metreveli. 1978. "New aspects of the sociology of disasters: A theoretical note." *Mass Emergencies* 3: 61–65.

Cochrane, H. 1975. *Natural Hazards and Their Distributive Effects.* Boulder, CO: University of Colorado, Institute of Behavioral Science, Program on Environment and Behavior.

Cochrane, H. 1997. "Forecasting the economic impact of a Midwest earthquake." Pp. 223–2247 in B. Jones (ed.) *Economic Consequences of Earthquakes: Preparing for the Unexpected.* Buffalo, NY: State University of New York at Buffalo, Multidisciplinary Center for Earthquake Engineering Research.

Cochrane, H. and J. Schmehl. 1993. "Financial repercussions: Implications for loss measurement." Pp. 161–189 in K. J. Tierney and J. M. Nigg (eds.) *Socioeconomic Impacts: Monograph No. 5, 1993 National Earthquake Conference.* Memphis: Central U.S. Earthquake Consortium.

Comfort, L. K. 1993. "Integrating information technology into international crisis management and policy." *Journal of Contingencies and Crisis Management* 1: 17–29.

Comfort, L. K. 1999. *Shared Risk: Complex Systems in Seismic Response*. Oxford, UK: Pergamon/Elsevier Science Ltd.

Cooke, D. 1995. "Los Angeles earthquake puts city disaster services to the test." *Disaster Recovery Journal* 7: 10–14.

Couch, S. R. and J. S. Kroll-Smith. 1985. "The chronic technological disaster: Toward a social scientific perspective." *Social Science Quarterly* 66: 564–575.

Covello, V. T., P. Slovic, and D. von Winterfeldt. 1987. *Risk Communication: A Review of the Literature*. Washington, DC: National Science Foundation.

Cuthbertson, B. and J. M. Nigg. 1987. "Technological disasters and the non-therapeutic community: A question of true victimization." *Environment and Behavior* 19: 462–483.

Cutter, S. L. 2001. *American Hazardscapes: The Regionalization of Hazards and Disasters*. Washington, DC: Joseph Henry Press.

Cutter, S. L., J. Tiefenbacher, and W. D. Solecki. 1992. "En-gendered fears: Femininity and technological risk perception." *Industrial Crisis Quarterly* 6: 5–22.

Dacy, D. C. and H. Kunreuther. 1969. *The Economics of Natural Disasters: Implications for Federal Policy*. New York: Free Press.

Dahlhamer, J. M. and L. Reshaur. 1996. "Businesses and the 1994 Northridge earthquake: An analysis of pre- and post-disaster preparedness." Newark, DE: University of Delaware, Disaster Research Center. Preliminary Paper No. 240.

Dahlhamer, J. M. and M. J. D'Souza. 1997. "Determinants of business disaster preparedness in two U.S. metropolitan areas." *International Journal of Mass Emergencies and Disasters* 15: 265–281.

Dash, N. 1997. "The use of geographic information systems in disaster research." *International Journal of Mass Emergencies and Disasters* 15: 135–146.

Dash, N., W. G. Peacock, and B. H. Morrow. 1997. "And the poor get poorer: A neglected Black community." Pp. 206–225 in W. G. Peacock, B. H. Morrow, and H. Gladwin (eds.) *Hurricane Andrew: Ethnicity, Gender and the Sociology of Disaster*. London: Routledge.

Davis, M. S. 1989. "Living along the fault line: An update on earthquake awareness and preparedness in Southern California." *Urban Resources* 5: 8–14.

De Man, A. and P. Simpson-Housley. 1987. "Factors in perception of earthquake hazard." *Perceptual and Motor Skills* 64: 815–820.

Devall, B. and G. Sessions. 1985. *Deep Ecology: Living as if Nature Mattered*. Salt Lake City: Peregrine Smith.

Disaster Information Task Force. 1997. *Harnessing Information and Technology for Disaster Management: The Global Disaster Information Network*. Washington, DC: Disaster Information Task Force.

Dohrenwend, B. P., B. S. Dohrenwend, G. J. Warheit, G. S. Bartlett, R. L. Goldsteen, K. Goldsteen, and J. L. Martin. 1981. "Stress in the community: A report to the President's Commission on the accident at Three Mile Island." Pp. 159–174 in T. H. Moss and D. L. Sills (eds.) *The Three Mile Island Nuclear Accident: Lessons and Implications*. New York: New York Academy of Sciences.

Dombrowsky, W. 1987. "Critical theory in disaster research." Pp. 331–356 in R. R. Dynes, B. De Marchi, and C. Pelanda (eds.) *Sociology of Disasters: Contributions of Sociology to Disaster Research*. Milan: Franco Angeli.

Dombrowsky, W. 1998. "Again and again: Is a disaster what we call a 'disaster'?" Pp. 19–30 in E. L. Quarantelli (ed.) *What Is a Disaster? Perspectives on the Question*. London: Routledge.

Dooley, D., R. Catalano, S. Mishra, and S. Serxner. 1992. "Earthquake preparedness: Predictors in a community survey." *Journal of Applied Social Psychology* 22: 451–470.

Drabek, T. E. 1965. Laboratory Simulation of a Police Communication System Under Stress. Doctoral Dissertation. Columbus, OH: Department of Sociology, Ohio State University.

Drabek, T. E. 1969. "Social processes in disaster: Family evacuation." *Social Problems* 16: 336–349.

Drabek, T. E. 1983a. "Shall we leave? A study on family reactions when disaster strikes." *Emergency Management Review* 1: 25–29.

Drabek, T. E. 1983b. "Alternative patterns of decision-making in emergent disaster response networks." *International Journal of Mass Emergencies and Disasters* 1: 277–305.

Drabek, T. E. 1985. "Managing the emergency response." *Public Administration Review* 45: 85–92.

Drabek, T. E. 1986. *Human System Responses to Disaster: An Inventory of Sociological Findings.* New York: Springer-Verlag.

Drabek, T. E. 1987. *The Professional Emergency Manager: Structures and Strategies for Success.* Boulder, CO: University of Colorado, Institute of Behavioral Science, Program on Environment and Behavior.

Drabek, T. E. 1989. "Disasters as non-routine social problems." *International Journal of Mass Emergencies and Disasters* 7: 253–264.

Drabek, T. E. 1990. *Emergency Management: Strategies for Maintaining Organizational Integrity.* New York: Springer-Verlag.

Drabek, T. E. 1991a. "Anticipating organizational evacuations: Disaster planning by managers of tourist-oriented private firms." *International Journal of Mass Emergencies and Disasters* 9: 219–245.

Drabek, T. E. 1991b. "The evolution of emergency management." Pp. 3–29 in T. E. Drabek and G. J. Hoetmer (eds.) *Emergency Management: Principles and Practice for Local Government.* Washington, DC: International City and County Management Association.

Drabek, T. E. 1991c. *Microcomputers in Emergency Management.* Boulder, BO: University of Colorado, Institute of Behavioral Science, Program on Environment and Behavior.

Drabek, T. E. 1993. "Major themes in disaster preparedness and response research." Paper presented at the Research Seminar on Socio-Economic Aspects of Disaster in Central America. San Jose, Costa Rica, January.

Drabek, T. E. 1994. *Disaster Evacuation and the Tourist Industry.* Boulder, CO: University of Colorado, Institute of Behavioral Science, Program on Environment and Behavior.

Drabek, T. E. 1995. "Disaster responses within the tourist industry." *International Journal of Mass Emergencies and Disasters* 13: 7–23.

Drabek, T. E. 1996. *Disaster Evacuation Behavior: Tourists and Other Transients.* Boulder, CO: University of Colorado, Institute of Behavioral Science, Program on Environment and Behavior.

Drabek, T. E. and J. E. Haas. 1969. "Laboratory simulation of organizational stress." *American Sociological Review* 34: 223–238.

Drabek, T. E. and K. S. Boggs. 1968. "Families in disaster: Reactions and relatives." *Journal of Marriage and the Family* 30: 443–451.

Drabek, T. E. and W. H. Key. 1984. *Conquering Disaster: Family Recovery and Long-Term Consequences.* New York: Irvington Publishers.

Drabek, T. E. and J. S. Stephenson, 1971. "When disaster strikes." *Journal of Applied Social Psychology* 1: 187–203.

Drabek, T. E., A. H. Mushkatel, and T. S. Kilijanek. 1983. *Earthquake Mitigation Policy: The Experience of Two States.* Boulder, CO: University of Colorado, Institute of Behavioral Science, Program on Environment and Behavior.

Drabek, T. E., H. L. Tamminga, T. S. Kilijanek, and C. R. Adams. 1982. *Managing Multiorganizational Emergency Responses: Emergent Search and Rescue Networks in Natural Disaster and Remote Area Settings.* Boulder, CO: University of Colorado, Institute of Behavioral Science, Program on Environment and Behavior.

Dunwoody, S. 1992. "The media and public perceptions of risk: How journalists frame risk stories." Pp. 75–100 in D. Bromley and K. Segerson (eds.) *The Social Response to Environmental Risk: Policy Formulation in an Age of Uncertainty.* Boston: Kluwer Academic Publishers.

Durham, T. and L. E. Suiter. 1991. "Perspectives and roles of the state and federal governments." Pp. 101–127 in T. E. Drabek and G. J. Hoetmer (eds.) *Emergency Management: Principles and Practice for Local Government.* Washington, DC: International City Management Association.

Durkin, M., S. Aroni, and A. Coulson. 1984. *Injuries in the Coalinga Earthquake of May 2, 1983.* Oakland, CA: Earthquake Engineering Research Institute.

Dynes, R. R. 1970. *Organized Behavior in Disaster.* Lexington, MA:Heath Lexington Books.

Dynes, R. R. 1993. "Disaster reduction: The importance of adequate assumptions about social organization." *Sociological Spectrum* 6: 24–25.

Dynes, R. R. 1994. "Community emergency planning: False assumptions and inappropriate analogies." *International Journal of Mass Emergencies and Disasters* 12: 141–158.

Dynes, R. R. 1998. "Coming to terms with community disaster." Pp. 109–126 in E. L. Quarantelli (ed.) *What Is a Disaster? Perspectives on the Question.* London: Routledge.

Dynes, R. R. and E. L. Quarantelli. 1968. "Group behavior under stress: A required convergence of organizational and collective behavior perspectives." *Sociology and Social Research* 52: 416–429.

Dynes, R. R. and E. L. Quarantelli. 1971. "The absence of community conflict in the early phases of natural disaster." Pp. 200–204 in C. G. Smith (ed.) *Conflict Resolution: Contributions of the Behavioral Sciences.* South Bend, IN: University of Notre Dame Press.

Dynes, R. R. and E. L. Quarantelli. 1976. "The family and community context of individual reactions to disaster." Pp. 231–245 in H. Parad, H. F. L. Resnick, and L. G. Parad (eds.) *Emergency and Disaster Management: A Mental Health Source Book.* Bowie, MD: The Charles Press.

Dynes, R. R. and E. L. Quarantelli. 1977a. *The Role of Local Civil Defense in Disaster Planning.* Newark, DE: University of Delaware, Disaster Research Center.

Dynes, R. R. and E. L. Quarantelli. 1977b. *Organizational Communications and Decision Making in Crisis.* Newark, DE: University of Delaware, Disaster Research Center.

Dynes, R. R. and K. J. Tierney. 1994. *Disasters, Collective Behavior, and Social Organization.* Newark, DE: University of Delaware Press.

Dynes, R. R., E. Haas, and E. L. Quarantelli. 1967. "Administrative, methodological, and theoretical problems of disaster research." *Indian Sociological Bulletin* 4: 215–227.

Dynes, R. R., E. L Quarantelli, and G. A. Kreps. 1981. *A Perspective on Disaster Planning.* Newark, DE: University of Delaware, Disaster Research Center (3rd edition; originally published in 1972).

Dynes, R. R., E. L. Quarantelli, and D. Wenger. 1990. *Individual and Organizational Response to the 1985 Earthquake in Mexico City, Mexico.* Newark, DE: University of Delaware, Disaster Research Center.

The Economist. 1997. "Japan: Volunteers step forward." April 12: 34.

Edwards, M. L. 1993. "Social location and self-protective behavior: Implications for earthquake preparedness." *International Journal of Mass Emergencies and Disaster* 11: 293–304.

Eguchi, R. T., H. A. Seligson, W. W. Hays, and L. S. Walter. 1998. "Global emergency risk management system (GERMS)—The integration of emerging technologies with disaster management." *Proceedings of the Eleventh European Conference on Earthquake Engineering.* Paris, France, September 6–11.

Enarson, E. 1998. "Through women's eyes: A gendered research agenda for disaster social science." *Disasters* 22: 157–173.

Enarson E. and B. H. Morrow. 1997. "A gendered perspective: The voices of women." Pp. 116–140 in Peacock. W. G., B. H. Morrow, and H. Gladwin (eds.) *Hurricane Andrew: Ethnicity, Gender and the Sociology of Disasters.* London: Routledge.

Enarson E. and B. H. Morrow (eds.). 1998. *The Gendered Terrain of Disaster.* Westport, CT: Praeger.

Erikson, K. T. 1994. *A New Species of Trouble: Explorations in Disaster, Trauma, and Community.* New York: W. W. Norton and Co.

Farley, J. E., H. D. Barlow, M. S. Finkelstein, and L. Riley. 1993. "Earthquake hysteria, before and after: A survey and follow-up on public response to the Browning forecast." *International Journal of Mass Emergencies and Disasters* 11: 305–322.

Faupel, C. E. and C. Bailey. 1988. "Contingencies affecting emergency preparedness for hazardous wastes." *International Journal of Mass Emergencies and Disasters* 6: 131–154.

Faupel, C. E., S. P. Kelley, and T. Petee. 1992. "The impact of disaster education on household preparedness for Hurricane Hugo." *International Journal of Mass Emergencies and Disasters* 10: 5–24.

Federal Emergency Management Agency. 1983. *Integrated Emergency Management System: Capability Assessment and Standards for State and Local Government: Interim Guidance.* Washington, DC: Federal Emergency Management Agency.

Federal Emergency Management Agency, 1987. *The California FIRESCOPE Program.* Emmitsburg, MD: National Emergency Training Center, Emergency Management Institute.

Federal Emergency Management Agency. 1993. *Improving Earthquake Mitigation: Report to Congress.* Washington, DC: Office of Earthquakes and Natural Hazards, Federal Emergency Management Agency.

Federal Emergency Management Agency. 1994. *Audit of FEMA's Comprehensive Cooperative Agreement Process.* Washington, DC: Office of Inspector General, Federal Emergency Management Agency.

Feldman, D. L. 1993. "SARA Title III and community hazards planning: The case of the chemical stockpile emergency preparedness program." *International Journal of Mass Emergencies and Disasters* 11: 85–97.

Feldman, J. and M. K. Lindell. 1990. "On rationality." Pp. 83–164 in I. Horowitz (ed.) *Recent Economic Thought: Organization and Decision Theory.* Boston: Kluwer Academic Publishers.

Fischer, H. W. III. 1998. *Response to Disaster: Fact Versus Fiction and Its Perpetuation: The Sociology of Disaster.* New York: University Press of America.

Fishbein, M. and I. Ajzen. 1975. *Belief, Attitude, Intention and Behavior: An Introduction to Theory and Research.* Reading, MA: Addison-Wesley.

Flynn, C. B. 1982. "Reactions of local residents to the accident at Three Mile Island." Pp. 49–61 in D. L. Sills, C. P. Wolf, and V. B. Shelanski (eds.) *Accident at Three Mile Island: The Human Dimensions.* Boulder, CO: Westview Press.

Flynn, J., P. Slovic, and C. K. Mertz. 1994. "Gender, race, and perception of environmental health risks." Risk Analysis 14: 1101–1108.

Ford, J. K. and A. Schmidt. 2000. "Emergency preparedness training: Strategies for enhancing real-world performance." *Journal of Hazardous Materials* 75: 195–215.

Fordham, M. 1998. "Making women visible in disasters: Problematising the private domain." *Disasters* 22: 126–143.

Fothergill, A. 1996. "Gender, risk, and disaster." *International Journal of Mass Emergencies and Disasters* 14: 33–56.

Fothergill, A. 1998. "The neglect of gender in disaster work: An overview of the literature." Pp. 11–25 in E. Enarson and B. H. Morrow (eds.) *The Gendered Terrain of Disaster*. Westport, CT: Praeger.

Freudenburg, W. R. 1997. "Contamination, corrosion, and the social order: An overview." *Current Sociology* 45: 19–39.

Freudenburg, W. R. and T. R. Jones. 1991. "Attitudes and stress in the presence of technological risk: A test of the Supreme Court hypothesis." *Social Forces* 69: 1143–1168.

Frey, R. S. 1995. "The international traffic in pesticides." *Technological Forecasting and Social Change* 50: 151–169.

Friedman, B. 1987. *The Art of Storytelling: The Structuring and Processing of News During Disasters*. Newark, DE: University of Delaware, Disaster Research Center.

Friedman, B., J. Linn, D. Lockwood, L. Snowden, and D. Zeidler. 1986. *Mass Media and Disaster: An Annotated Bibliography*. Newark, DE: University of Delaware, Disaster Research Center. Miscellaneous Report No. 36.

Friedman, S. M. 1989. "TMI: The media story that will not die." Pp. 63–83 in L. Masel Walters, L. Wilkins, and T. Walters (eds.) *Bad Tidings: Communication and Catastrophe*. Hillsdale, NJ: Lawrence Erlbaum Associates.

Friedsam, H. J. 1962. "Older persons in disaster." Pp. 151–182 in G. W. Baker and D. W. Chapman (eds.) *Man and Society in Disaster*. New York: Basic Books.

Friesema, H. P., J. Caparano, G. Goldstein, R. Lineberry, and R. McCleary. 1979. *Aftermath: Communities After Natural Disasters*. Thousand Oaks, CA: Sage Publications.

Fritz, C. E. 1961a. "Disasters." Pp. 651–694 in R. K. Merton and R. A. Nisbet (eds.) *Contemporary Social Problems*. New York: Harcourt.

Fritz, C. E. 1961b. Published as Fritz, C. E. 1996. *Disasters and Mental Health: Therapeutic Principles Drawn from Disaster Studies*. Newark, DE: University of Delaware, Disaster Research Center. Historical and Comparative Disaster Series No. 10.

Fritz, C. E. and E. Marks. 1954. "The NORC studies of human behavior in disaster." *Journal of Social Issues* 10: 26–41.

Fritz, C. E. and J. H. Mathewson. 1957. *Convergence Behavior in Disasters*. Washington, DC: National Academy of Sciences/National Research Council.

Gaard, G. 1998. *Ecological Politics: Ecofeminism and the Greens*. Philadelphia: Temple University Press.

Gabor, T. 1981. "Mutual aid systems in the United States for chemical emergencies." *Journal of Hazardous Materials* 4: 343–356.

Gamson. W. A. and A. Modigliani. 1989. "Media discourse and public opinion on nuclear power." *American Journal of Sociology* 95: 1–37.

Gans, Herbert. 1980. Deciding What's News. New York: Vintage Books.

Geis, D. E. 1996. "Creating sustainable and disaster-resistant communities." Paper presented at the Aspen Global Change Institute, Aspen, CO, July 10.

Giddens, A. 1984. *The Constitution of Society: Outline of the Theory of Structuration*. Berkeley: University of California Press.

Gilbert, C. 1998. "Studying disaster: Changes in the main conceptual tools." Pp. 11–18 in E. L. Quarantelli (ed.) *What is a Disaster? Perspectives on the Question*. London: Routledge.

Gillespie, D. F. 1991. "Coordinating community resources." Pp. 55–78 in T. E. Drabek and G. J. Hoetmer (eds.) *Emergency Management: Principles and Practice for Local Government*. Washington, DC: International City Management Association.

Gillespie, D. F. and C. L. Streeter. 1987. "Conceptualizing and measuring disaster preparedness." *International Journal of Mass Emergencies and Disasters* 5: 155–176.

Gillespie, D. F., D. S. Mileti, and R. Perry. 1976. *Organizational Response to Changing Community Systems*. Kent, OH: Kent State University Press.

Gillespie, D. F., R. A. Colignon, M. M. Banerjee, S. A. Murty, and M. Rogge. 1992. *Interorganizational Relations for Disaster Preparedness*. St. Louis, MO: George Warren Brown School of Social Work. Final report submitted to the National Science Foundation.

Gillespie, D. F., R. A. Colignon, M. M. Banerjee, S. A. Murty, and M. Rogge. 1993. *Partnerships for Community Preparedness*. Boulder, CO: University of Colorado, Institute of Behavioral Science, Program on Environment and Behavior. Monograph No. 54.

Gillespie, D. F., S. A. Murty, M. E. Rogge, K. J. Robards, and C.-Y. Shen. 1995. *Assessment of the FEMA Earthquake Hazard Reduction Program. Final report submitted to the Federal Emergency Management Agency*. St. Louis, MO: Washington University.

Gitlin, T. 1980. *The Whole World Is Watching: Mass Media in the Making and Unmaking of the New Left*. Berkeley: University of California Press.

Gladwin, H. and W. G. Peacock. 1997. "Warning and evacuation: A night for hard houses." Pp. 52–74 in W. G. Peacock, B. H. Morrow, and H. Gladwin (eds.) *Hurricane Andrew: Ethnicity, Gender and the Sociology of Disasters*. London: Routledge.

Goltz, J. D. 1985. "Are the news media responsible for disaster myths? A content analysis of emergency response imagery." Los Angeles: Department of Sociology, University of California, Los Angeles.

Goltz, J. D., L. A. Russell, and L. B. Bourque. 1992. "Initial behavioral response to a rapid onset disaster: A case study of the October 1, 1987 Whittier Narrows earthquake." *International Journal of Mass Emergencies and Disasters* 10: 43–69.

Goode, E. 1992. *Collective Behavior*. Fort Worth: Saunders College Publishing/Harcourt Brace Jovanovich.

Gramling, R. and W. R. Freudenburg. 1992. "The Exxon Valdez spill in the context of US petroleum politics." *Industrial Crisis Quarterly* 6: 175–196.

Gray, J. K. 1981. "Characteristic patterns of and variations in community response to acute chemical emergencies." *Journal of Hazardous Materials* 4: 357–366.

Green, B. L., M. Korol, M. C. Grace, M. G. Vary, A. C. Leonard, G. C. Gleser, and S. Smitson-Cohen. 1991. "Children and disaster: Age, gender, and parental effects on PTSD symptoms." *Journal of the American Academy of Child and Adolescent Psychiatry* 30: 945–951.

Gruntfest, E. and M. Weber. 1998. "Internet and emergency management: Prospects for the future." *International Journal of Mass Emergencies and Disasters* 16: 55–72.

Gulaid, J. A., Sacks, J. J., and Sattin, R. W. 1989. "Deaths from residential fires among older people, United States, 1984." *Journal of the American Geriatric Society* 37: 331–334.

Haas, J. E., H. Cochrane, and D. Eddy. 1977. "Consequences of a cyclone for a small city." *Ekistics* 44: 45–51.

Hagan, J. and R. D. Peterson (eds.). 1995. *Crime and Inequality*. Stanford, CA: Stanford University Press.

Hamilton, L. C. 1985. "Concern about toxic wastes: Three demographic predictors." *Sociological Perspectives* 28: 263–286.

Hannigan, J. A. 1995. *Environmental Sociology: A Social Constructionist Approach*. London: Routledge.

Hannigan, J. A. and R. M. Kueneman. 1978. "Anticipating flood emergencies: A case study of a Canadian disaster subculture." Pp. 129–146 in E. L. Quarantelli (ed.) *Disasters: Theory and Research.* Thousand Oaks, CA: Sage Publications.

Hans, J. and T. Sell. 1974. *Evacuation Risks—An Evaluation.* Las Vegas: Environmental Protection Agency, National Environmental Research Center.

Harrald, J. R., R. Cohn, and W. A. Wallace. 1992. "'We were always re-organizing . . . ' Some crisis management implications of the *Exxon Valdez.*" *Industrial Crisis Quarterly* 6: 197–217.

Healy, R. J. 1969. *Emergency and Disaster Planning.* New York: Wiley.

Hewitt, K. (ed.) 1983. *Interpretations of Calamity: From the Viewpoint of Human Ecology.* London: Allen and Unwin.

Hilgartner, S. 1992. "The social construction of risk objects." Pp. 40–53 in J. F. Short and L. B. Clarke (eds.) *Organizations, Uncertainties, and Risk.* Boulder, CO: Westview Press.

Hill, R. and D. Hanson. 1962. "Families in disaster." Pp. 185–221 In G. W. Baker and D. Chapman (eds.) *Man and Society in Disaster.* New York: Basic Books.

Hoetmer, G. J. 1983. *Emergency Management: Individual and County Data.* Washington, DC: International City Management Association.

Holton, J. L. 1985. *The Electronic Media and Disasters in the High-Tech Age.* Washington, DC: Federal Emergency Management Agency.

Horlick-Jones, T. 1995. "Modern disasters as outrage and betrayal." *International Journal of Mass Emergencies and Disasters* 13:305–316.

Houts, P. S., P. D. Cleary, and T.-W. Hu. 1988. *The Three Mile Island Crisis.* State College, PA: Penn State University Press.

Houts, P. S., M. K. Lindell, T.-W. Hu, P. D. Cleary, G. Tokuhata, and C. B. Flynn. 1984. "The protective action decision model applied to evacuation during the Three Mile Island crisis." *International Journal of Mass Emergencies and Disasters* 2: 27–39.

Huerta, F. and R. Horton. 1978. "Coping behavior of elderly flood victims." *Gerontologist* 18: 541–546.

International Federation of Red Cross and Red Crescent Societies. 1993. *World Disaster Report, 1993.* Dordrecht, The Netherlands: Kluwer Academic Publishers.

Ives, S. and O. Furuseth. 1980. "Immediate Response to Headwater Flooding in Neighborhoods in Charlotte, North Carolina." Charlotte: Department of Sociology, University of North Carolina.

Jackson, E. L. 1977. "Public response to earthquake hazard." California Geology 30: 278–280.

Jackson, E. L. 1981. "Response to earthquake hazard: The West Coast of North America." *Environment and Behavior* 13: 387–416.

Jackson, E. L. and T. Mukerjee. 1974. "Human adjustment to the earthquake hazard of San Francisco, CA." Pp. 160–166 in Gilbert White (ed.) *Natural Hazards: Local, National, Global.* New York: Oxford University Press.

James, T. F. and D. E. Wenger. 1980. "Public perceptions of disaster-related behaviors." Pp. 162–166 in E. J. Baker (ed.) *Hurricanes and Coastal Storms: Awareness, Evacuation and Mitigation.* Tallahassee, FL: Florida State University.

Jasper, J. M. 1990. *Nuclear Politics: Energy and the State in the United States, Sweden, and France.* Princeton, NJ: Princeton University Press.

Johnson, N. R. 1987. "Panic at 'the Who concert stampede': An empirical assessment." *Social Problems* 34: 362–373.

Johnson, N. R. 1988. "Fire in a crowded theater: A descriptive analysis of the emergence of panic." *International Journal of Mass Emergencies and Disasters* 6: 7–26.

Johnson, N. R., W. E. Feinberg, and D. M. Johnston. 1994. "Microstructure and panic: The impact of social bonds on individual action in collective flight from the Beverly

Hills Supper Club fire." Pp. 168–189 in R. R. Dynes and K. J. Tierney (eds.) *Disasters, Collective Behavior, and Social Organization.* Newark, DE: University of Delaware Press.

Johnson, N. R., D. M. Johnston, and J. E. Peters. 1989. "At a competitive disadvantage? The fate of the elderly in collective flight." Paper presented at the annual meeting of the North Central Sociological Association, Akron, OH.

Johnston, D. M. and N. R. Johnson. 1989. "Role expansion in disaster: An investigation of employee behavior in a nightclub fire." *Sociological Focus* 22: 39–51.

Kaniasty, K. and F. H. Norris. 1993. "A test of the social support deterioration model in the context of natural disaster." *Journal of Personality and Social Psychology* 64: 395–408.

Kartez, J. D. 1984. "Crisis response planning: Toward a contingent analysis." *Journal of the American Planning Association* 50: 9–21.

Kartez, J. D. and W. J. Kelley. 1988. "Research-based disaster planning: Conditions for implementation." Pp. 126–146 in L. K. Comfort (ed.) *Managing Disaster: Strategies and Policy Perspectives.* Durham, NC: Duke University Press.

Kartez, J. D. and M. K. Lindell. 1987. "Planning for uncertainty: The case of local disaster planning." *American Planning Association Journal* 53: 487–498.

Kartez, J. D. and M. K. Lindell. 1990. "Adaptive planning for community disaster response." Pp. 5–31 in R. T. Sylves and W. L. Waugh (eds.) *Cities and Disaster: North American Studies in Emergency Management.* Springfield, IL: Charles C Thomas.

Kasperson, R. E. and K. D. Pijawka. 1985. "Societal response to hazards and major hazard events: Comparing natural and technological hazards." *Public Administration Review* 45: 7–18.

Kates, R. W. 1962. "Hazard and choice perception in flood plain management." Chicago: University of Chicago, Department of Geography. Research Paper No. 78.

Keating, J. P., Elizabeth F. Loftus, and M. Manber. 1983. "Emergency evacuations during fires: Psychological considerations." Pp. 83–99 in R. F. Kidd and M. J. Saks (eds.) *Advances in Applied Social Psychology.* Hillsdale, NJ: Lawrence Erlbaum Associations.

Kessler, R. C. and H. W. Neighbors. 1986. "A new perspective on the relationships among race, social class, and psychological distress." *Journal of Health and Social Behavior* 27: 107–115.

Kessler, R. C., J. B. Turner, and J. S. House. 1989. "Unemployment, reimployment, and emotional functioning in a community sample." *American Sociological Review* 54: 648–657.

Killian, L. M. 1952. "The significance of multi-group membership in disasters." *American Journal of Sociology* 57: 309–314.

Knoke, D. 1990. *Political Networks: The Structural Perspective.* Cambridge and New York: Cambridge University Press.

Kreps, G. A. 1984. "Sociological inquiry and disaster research." *Annual Review of Sociology* 10: 309–330.

Kreps, G. A. 1985. "Disaster and the social order." *Sociological Theory* 3: 49–65.

Kreps, G. A. (ed.) 1989. *Social Structure and Disaster.* Newark, DE: University of Delaware Press.

Kreps, G. A. 1990. "The federal emergency management system in the United States: Past and present." Paper presented at the 12th World Congress of Sociology, Madrid, Spain, July 9–13.

Kreps, G. A. 1991. "Organizing for emergency management." Pp. 30–54 in T. E. Drabek and G. J. Hoetmer (eds.) *Emergency Management: Principles and Practice for Local Government.* Washington, DC: International City Management Association.

Kreps, G. A. 1998. "Disaster as systemic event and social catalyst." Pp. 31–55 in E. L. Quarantelli (ed.) *What Is a Disaster? Perspectives on the Question.* London: Routledge.

Kreps, G. A. and S. L. Bosworth. 1993. "Disaster, organizing, and role enactment: A structural approach." *American Journal of Sociology* 99: 428–463.

Kreps, G. A. and T. E. Drabek. 1996. "Disasters are nonroutine social problems." *International Journal of Mass Emergencies and Disasters* 14: 129–153.

Krieg, E. J. 1995. "A sociohistorical interpretation of toxic waste sites: The case of Greater Boston." *American Journal of Economics and Sociology* 54: 1–14.

Krieg, E. J. 1998. "The two faces of toxic waste: Trends in the spread of environmental hazards." *Sociological Forum* 13: 3–20.

Kroll-Smith, J. S. and S. R. Couch. 1990. *The Real Disaster Is Above Ground.* Lexington: University Press of Kentucky.

Kroll-Smith, J. S. and S. R. Couch. 1991. "What is a disaster? An ecological-symbolic approach to resolving the definitional debate." *International Journal of Mass Emergencies and Disasters* 9: 355–366.

Kunreuther, H. 1992. "A conceptual framework for managing low-probability events." Pp. 301–320 in S. Krimsky and D. Golding (eds.) *Social Theories of Risk.* Westport, CT: Praeger.

Kunreuther, H. and R. J. Roth, Sr. (eds.) 1998. *Paying the Price: The Status and Role of Insurance Against Natural Hazards in the United States.* Washington, DC: Joseph Henry Press.

Kunreuther, H., R. Ginsburg, L. Miller, P. Sagi, P. Slovic, B. Borkan, and N. Katz. 1978. *Disaster Insurance Protection: Public Policy Lessons.* New York: John Wiley and Sons.

Labadie, J. 1984. "Problems in local emergency management." *Environmental Management* 8: 489–494.

Lachman, R., Tatsuoka, M. and W. Bonk. 1961. "Human behavior during the tsunami of May, 1960." *Science* 133: 1405–1409.

Lambright, W. H. 1985. "The Southern California Earthquake Preparedness Project: Evolution of an 'earthquake entrepreneur.'" *International Journal of Mass Emergencies and Disasters* 3L 75–94.

Landesman, L. Y. 1989. "Improving medical preparedness for chemical accidents: An interorganizational resource review." *International Journal of Mass Emergencies and Disasters* 7: 152–166.

Landesman, L. Y. (ed.) 1996. *Emergency Preparedness in Health Care Organizations.* Oakbrook Terrace, IL: Joint Commission on Accreditation of Healthcare Organizations.

La Porte, T. R. 1988. "The United States air traffic control system: Increasing reliability in the midst of rapid growth." Pp. 215–244 in T. Hughes and R. Mayntz (eds.) *The Development of Large Scale Technical Systems.* Boulder, CO: Westview Press.

La Porte, T. R. and P. M. Consolini. 1991. "Working in practice but not in theory: Theoretical challenges of high reliability organizations." *Journal of Public Administration Research and Theory* 1: 19–47.

Lasswell, H. D. 1948. "The structure and function of communication in society." Pp. 37–51 in L. Bryson (ed.) *Communication of Ideas.* New York: Harper.

Ledingham, J. A. and L. Masel Walters. 1989. "The sound and the fury: Mass media and hurricanes." Pp. 35–45 in L. M. Walters, L. Wilkins, and T. Walters (eds.) *Bad Tidings: Communication and Catastrophe.* Hillsdale, NJ: Lawrence Erlbaum Associates.

Lindell, M. K. 1994a. "Are local emergency planning committees effective in developing community disaster preparedness?" *International Journal of Mass Emergencies and Disasters* 12: 159–182.

Lindell, M. K. 1994b. "Perceived characteristics of environmental hazards." *International Journal of Mass Emergencies and Disasters* 12: 303–326.

Lindell, M. K. 1995. "Assessing emergency preparedness in support of hazardous facility risk analyses: An application at a U.S. hazardous waste incinerator." *Journal of Hazardous Materials* 40: 297–193.

Lindell, M. K. and V. Barnes. 1986. "Protective response to technological emergency." *Nuclear Safety* 27: 457–467.

Lindell, M. K. and M. Meier. 1994. "Planning effectiveness: Effectiveness of community planning for toxic chemical emergencies." *Journal of the American Planning Association* 60: 222–234.

Lindell, M. K. and R. W. Perry. 1980. "Evaluation criteria for emergency response plans in radiological transportation." *Journal of Hazardous Materials* 3: 335–348.

Lindell, M. K. and R. W. Perry. 1983. "Nuclear power plant emergency warning: How would the public respond?" *Nuclear News* 26: 49–53.

Lindell, M. K. and R. W. Perry. 1987. "Warning mechanisms in emergency response systems" *International Journal of Mass Emergencies and Disasters* 5: 137–153.

Lindell, M. K. and R. W. Perry. 1992. *Behavioral Foundations of Community Emergency Management.* Washington, DC: Hemisphere Publishing Corporation.

Lindell, M. K. and R. W. Perry. 1996. "Assessing gaps in environmental emergency planning." *Journal of Environmental Planning and Management* 39: 541–545.

Lindell, M. K. and R. W. Perry. 1998. "Earthquake impacts and hazard adjustments by acutely hazardous materials facilities following the Northridge earthquake." *Earthquake Spectra* 14: 285–299.

Lindell, M. K. and R. W. Perry. 2000. "Household adjustment to earthquake hazard: A review of research." *Environment and Behavior* 32: 590–630.

Lindell, M. K. and C. S. Prater. 2000. "Household adoption of seismic hazard adjustments: A comparison of residents in two states." *International Journal of Mass Emergencies and Disasters* 18: 317–338.

Lindell, M. K. and D. J. Whitney. 1995. "Effects of organizational environment, internal structure and team climate on the effectiveness of local emergency planning committees." *Risk Analysis* 15: 439–447.

Lindell, M. K. and D. J. Whitney. 2000. "Correlates of seismic hazard adjustment adoption." *Risk Analysis* 20:13–25.

Lindell, M. K., D. J. Whitney, C. J. Futch, and C. S. Clause. 1996a. "Multi-method assessment of organizational effectiveness in a local emergency planning committee." *International Journal of Mass Emergencies and Disasters* 14: 195–220.

Lindell, M. K., D. J. Whitney, C. J. Futch, and C. S. Clause. 1996b. "The local emergency planning committee: A better way to coordinate disaster planning." Pp. 234–249 in R. T. Sylves and W. L. Waugh, Jr. (eds.) *Disaster Management in the U.S. and Canada,* 2nd edition. Springfield, IL: Charles C Thomas.

Lindell, M. K., P. A. Bolton, R. W. Perry, G. A. Stoetzel, J. B. Martin, and C. B. Flynn. 1985. *Planning concepts and decision criteria for sheltering and evacuation in a nuclear power plant emergency.* AIF/NESP-031. Bethesda, MD: Atomic Industrial Forum/National Environmental Studies Project.

Lindell, M. K., with D. Alesch, P. A. Bolton. M. R. Greene, L. A. Larson, R. Lopes, P. J. May, J.-P. Mulilis, S. Nathe, J. M. Nigg, R. Palm, P. Pate, R. W. Perry, J. Pine, S. K. Tubbesing, and D. J. Whitney. 1997. "Adoption and implementation of hazard adjustments." *International Journal of Mass Emergencies and Disasters*, Special Issue 15: 327–453.

Logan, J. R. and H. L. Molotch. 1987. *Urban Fortunes: The Political Economy of Place.* Berkeley: University of California Press.

Luhmann, N. 1993. *Risk: A Sociological Theory.* New York: Walter de Gruyter.

March, J. G. and J. Olsen. 1979. *Ambiguity and Choice in Organizations.* Bergen, Norway: Universitetsforlanget.

Marsden, P. V. and N. Lin. 1982. *Social Structure and Network Analysis.* Thousand Oaks, CA: Sage Publications.

Marston, S. A. (ed.) 1986. *Terminal Disasters: Computer Applications in Emergency Management.* Boulder, CO: University of Colorado, Institute of Behavioral Science, Program on Environment and Behavior.

Martinez, R. 1996. "Latinos and lethal violence: The impact of poverty and inequality." *Social Problems* 43: 131–146.

Marx, G. T. and D. McAdam. 1994. *Collective Behavior and Social Movements: Process and Structure.* Englewood Cliffs, NJ: Prentice Hall.

Maskrey, A. 1994. "Disaster mitigation as a crisis of paradigms: Reconstructing after the Alto Mayo earthquake, Peru." Pp. 109–123 in A. Varley (ed.) *Disasters, Development and Environment.* New York: John Wiley and Sons.

May, P. J. and R. E. Deyle. 1998. "Governing land use in hazardous areas with a patchwork system." Pp. 57–82 in R. Burby (ed.) *Cooperating With Nature: Confronting Natural Hazards with Land-Use Planning for Sustainable Communities.* Washington, DC: Joseph Henry Press.

May, P. J. and W. Williams. 1986. Disaster Policy Implementation: Managing Programs Under Shared Governance. New York: Plenum Press.

McLeod, J. D. and R. C. Kessler. 1990. "Socioeconomic status differences in vulnerability to undesirable life events." *Journal of Health and Social Behavior* 31: 162–172.

McLoughlin, D. 1985. "A framework for integrated emergency management." *Public Administration Review* 45: 165–172.

McLuckie, B. F. 1970. *Warning Systems in Disaster.* Newark, DE: University of Delaware, Disaster Research Center.

McLuckie, B. F. 1977. *Italy, Japan, and the United States: Effects of Centralization on Disaster Response 1964–1969.* Newark, DE: University of Delaware, Disaster Research Center.

Melick, M. E. and J. N. Logue. 1985. "The effect of disaster on the health and well-being of older women." *International Journal of Aging and Human Development* 21: 27–38.

Meyer, A. and J. Goes. 1989. "Organizational assimilation of innovations." *Academy of Management Journal* 31: 897–923.

Mileti, D. S. 1975a. *Natural Hazards Warning Systems in the United States.* Boulder, CO: University of Colorado, Institute of Behavioral Science, Program on Environment and Behavior.

Mileti, D. S. 1975b. *Disaster Relief and Rehabilitation in the United States: A Research Assessment.* Boulder, CO: University of Colorado, Institute of Behavioral Science, Program on Environment and Behavior.

Mileti, D. S. 1983. "Societal comparisons of organizational response to earthquake predictions." *International Journal of Mass Emergencies and Disasters* 1: 399–413.

Mileti, D. S. 1999. *Disasters by Design: A Reassessment of Natural Hazards in the United States.* Washington, DC: Joseph Henry Press.

Mileti, D. S. and J. D. Darlington. 1995. "Societal response to revised earthquake probabilities in the San Francisco Bay Area." *International Journal of Mass Emergencies and Disasters* 13: 119–145.

Mileti, D. S. and J. D. Darlington. 1997. "The role of searching in shaping reactions to earthquake risk information." *Social Problems* 44: 89–103.

Mileti, D. S. and C. Fitzpatrick. 1993. *The Great Earthquake Experiment: Risk Communication and Public Action.* Boulder, CO: Westview Press.

Mileti, D. S. and P. O'Brien. 1992. "Warnings during disaster: Normalizing communicated risk." *Social Problems* 39: 40–57.

Mileti, D. S. and J. Sorensen. 1987. "Determinants of organizational effectiveness in responding to low probability catastrophic events." *Columbia Journal of World Business* 22: 92–101.

Mileti, D. S., T. E. Drabek, and J. E. Haas. 1975. *Human Systems in Extreme Environments.* Boulder, CO: University of Colorado, Institute of Behavioral Science, Program on Environment and Behavior.

Mileti, D. S., D. F. Gillespie, and R. W. Perry. 1975. "The analytic use of case study materials." *Sociological Inquiry* 45: 70–75.

Mileti, D. S., C. Fitzpatrick, and B. C. Farhar. 1990. *Risk Communication and Public Response to the Parkfield Earthquake Prediction Experiment.* Fort Collins, CO: Hazards Assessment Laboratory and Department of Sociology, Colorado State University.

Mileti, D. S., J. H. Sorensen, and P. W. O'Brien. 1992. "Toward an explanation of mass care shelter use in evacuations." *International Journal of Mass Emergencies and Disasters* 10: 25–42.

Mileti, D. S., J. D. Darlington, C. Fitzpatrick, and P. W. O'Brien. 1993. *Communicating Earthquake Risk: Societal Response to Revised Probabilities in the Bay Area.* Fort Collins, CO: Hazards Assessment Laboratory and Department of Sociology, Colorado State University.

Miller, G. and J. A. Holstein (eds.) 1993. *Constructivist Controversies: Issues in Social Problems Theory.* New York: Aldine de Gruyter.

Mittler, E. 1989. *Natural Hazard Policy Setting: Identifying Supporters and Opponents of Nonstructural Hazard Mitigation.* Boulder, CO: University of Colorado, Institute of Behavioral Science, Program on Environment and Behavior.

Mitroff, I. I. and C. M. Pearson. 1993. *Coping With the Unthinkable.* New York: Jossey-Bass.

Mitroff, I. I., T. Pauchant, and P. Shrivastava. 1988. "The structure of man-made organizational crisis: conceptual and empirical issues in the development of a general theory of crisis management." *Technological Forecasting and Social Change* 33: 83–107.

Mitroff, I. I., T. Pauchant, M. Finney, and C. Pearson. 1989. "Do (some) organizations cause their own crises? Cultural profiles of crisis prone versus crisis prepared organizations." *Industrial Crisis Quarterly* 3: 269–283.

Mogil, H. M. 1980. "The weather warning and preparedness programs: The National Weather Service and the mass media." Pp. 205–213 in *Disasters and the Mass Media: Proceedings of the Committee on Disasters and the Mass Media Workshop.* Washington, DC: National Academy of Sciences/National Research Council.

Moore, H. E. 1964. *. . . .And the Winds Blew.* Austin: University of Texas Press.

Moore, H. E., F. Bates, M. Lyman, and V. Parenton. 1963. *Before the Wind: A Study of Repsonse to Hurricane Carla.* Washington, DC: National Academy of Sciences/National Research Council.

Morrow, B. H. and E. Enarson. 1994. "Making the case for gendered disaster research." Paper presented at the 13th World Congress of Sociology, Bielefeld, Germany, July.

Morrow, B. H. and E. Enarson. 1996. "Hurricane Andrew through women's eyes: Issues and recommendations." *International Journal of Mass Emergencies and Disasters* 14: 5–22.

Mulford, C. L., G. E. Klonglan, and J. P. Kopachevsky. 1973. "Securing Community Resources for Social Action." Ames, IA: Iowa State University, Department of Sociology and Anthropology.

Mulilis, J.-P. and R. A. Lippa. 1990. "Behavioral change in earthquake preparedness due to negative threat appeals: A test of protection motivation theory." *Journal of Applied Social Psychology* 20: 619–638.

Mulilis, J.-P. and T. S. Duval. 1995. "Negative threat appeals and earthquake prepared-ness: A person-relative-to-event (PrE) model of coping with threat." *Journal of Applied Social Psychology* 25: 1319–1339.

Murphy, R. 1994. *Rationality and Nature: A Sociological Inquiry into a Changing Relationship.* Boulder, CO: Westview Press.

Murrell, S. A. and F. H. Norris. 1984. "Resources, life events, and changes in positive affect and depression in older adults." *American Journal of Community Psychology* 12: 445–464.

Mushkatel, A. H. and L. F. Weschler. 1985. "Emergency management and the intergovern-mental system." *Public Administration Review* 45: 49–56.

National Academy of Public Administration. 1993. *Coping With Catastrophe: Building an Emergency Management System to Meet People's Needs in Natural and Manmade Disasters.* Washington, DC: National Academy of Public Administration.

National Academy of Sciences. 1980. *Disasters and the Mass Media: Proceedings of the Committee on Disasters and the Mass Media Workshop, February, 1979.* Committee on Disasters and the Mass Media, National Research Council. Washington, DC: National Academy of Sciences.

National EMS Clearinghouse. 1988. *State EMS Office: Communications Programs and Disaster Preparedness.* Lexington, KY: National EMS Clearinghouse and Charles McC Mathias National Study Center for Trauma and EMS.

National Governors' Association. 1978a. *Federal Emergency Authorities: Abstracts.* Washington, DC: National Governors' Association.

National Governors' Association. 1978b. *National Emergency Assistance Programs: A Governor's Guide.* Washington, DC: National Governors' Association.

National Governors' Association. 1978c. *Emergency Preparedness Project: Final Report.* Washington, DC: National Governors' Association.

National Governors' Association. 1979. *Comprehensive Emergency Management: A Governor's Guide.* Washington, DC: National Governors' Association.

Neal, D. M. 1983. "A structural analysis of the emergence and non-emergence of citizens' groups in disaster threat situations." Newark, DE: University of Delaware, Disaster Research Center.

Neal, D. M. 1984. "Blame assignment in a diffuse disaster situation: A case example of the role of an emergent citizen group." *International Journal of Mass Emergencies and Disasters* 2: 251–266.

Neal, D. M. 1994. "The consequences of excessive unrequested donations: The case of Hurricane Andrew." *Disaster Management* 6: 23–28.

Neal, D. M. and B. D. Phillips. 1990. "Female-dominated local social movement organiza-tions in disaster-threat situations." Pp. 243–255 in G. West and R. L. Blumberg (eds.) *Women and Social Protest.* New York: Oxford University Press.

Neal, D. M. and B. D. Phillips. 1995. "Effective emergency management: Reconsidering the bureaucratic approach." *Disasters* 19: 327–337.

Neal, D. M., J. B. Perry, R. Hawkins, and K. Green. 1988. "Patterns of giving and receiv-ing help during severe winter conditions." *Disasters* 12: 77–85.

Nelson, L. S. and R. W. Perry. 1991. "Organizing public education for technological emer-gencies." *Disaster Management* 4: 21–26.

Nigg, J. M. 1982. "Communication under conditions of uncertainty: Understanding earth-quake forecasting." *Journal of Communication* 32: 27–36.

Nigg, J. M. 1987. "Communication and behavior: Organizational and individual response to warnings." Pp. 103–117 in R. R. Dynes, B. De Marchi, and C. Pelanda (eds.) *Sociology of Disasters: Contribution of Sociology to Disaster Research.* Milan: Franco Angeli.

Nigg, J. M. 1997. "Emergency response following the 1994 Northridge earthquake: Inter-governmental coordination issues." Newark, DE: University of Delaware Disaster Research Center. Preliminary Paper No. 250.

Nimmo, D. 1984. "TV network news coverage of Three Mile Island: Reporting disasters as technological fables." *International Journal of Mass Emergencies and Disasters* 2: 115–145.

Noji, E. K. 1989. "The 1988 earthquake in Soviet Armenia: Implications for earthquake preparedness." *Disasters* 13: 255–262.

Noji, E. K. 1995. "Natural disaster management." Pp. 644–663 in Auerbach, P. S. and E. C. Geehr (eds.) *Management of Wilderness and Environmental Emergencies*, 3rd edition. St. Louis: C.V. Mosby.

Noji, E. K. 1997. *The Public Health Consequences of Disasters*. New York: Oxford University Press.

Norris, F. H. 1992. "Epidemiology of trauma: Frequency and impact of different potentially traumatic events on different demographic groups." *Journal of Consulting and Clinical Psychology* 60: 409–418.

O'Brien, P. and D. S. Mileti. 1992. "Citizen participation in emergency response following the Loma Prieta earthquake." *International Journal of Mass Emergencies and Disasters* 10: 71–89.

Oliver-Smith, A. 1986. *The Martyred City: Death and Rebirth in the Andes*. Albuquerque: University of New Mexico Press.

Oliver-Smith, A. 1998. "Global changes and definitions of disaster." Pp. 177–194 in E. L. Quarantelli (ed.) *What Is a Disaster? Perspectives on the Question*. London: Routledge.

Oliver-Smith, A. 1990. "Post-disaster housing reconstruction and social inequality: A challenge to policy and practice." *Disasters* 14: 7–19.

Oliver-Smith, A. 1994. "Peru's five hundred year earthquake: Vulnerability in historical context." Pp. 31–48 in A. Varley (ed.) *Disasters, Development, and Environment*. New York: John Wiley and Sons.

Olshansky, R. 1994. "Seismic hazard mitigation in the Central United States: The role of states." Pp. G1–G16 in Shedlock, K. M. and A. C. Johnston (eds.) *Investigations of the New Madrid Seismic Zone*. Reston, VA: United States Geological Survey.

Pace, S., K. O'Connell, and B. Lachman. 1997. *Using Intelligence Data for Environmental Needs*. Santa Monica, CA: RAND Corporation.

Palm, R. 1990. *Natural Hazards: An Integrative Framework for Research and Planning*. Baltimore, MD: Johns Hopkins University Press.

Palm, R. and M. E. Hodgson. 1992. *After a California Earthquake: Attitude and Behavior Change*. Chicago: University of Chicago Press.

Palm, R., M. E. Hodgson, R. D. Blanchard, and D. Lyons. 1990. *Earthquake Insurance in California*. Boulder, CO: Westview Press.

Parr, A. R. 1970. "Organizational response to community crises and group emergence." *American Behavioral Scientist* 13: 424–427.

Patterson, P. 1989. "Reporting Chernobyl: Cutting the government fog to cover the nuclear cloud." Pp. 131–147 in L. M. Walters, L. Wilkins, and T. Walters (eds.) *Bad Tidings: Communication and Catastrophe*. Hillsdale, NJ: Lawrence Erlbaum Associates.

Peacock, W. G., B. H. Morrow, and H. Gladwin (eds.). 1997. *Hurricane Andrew: Ethnicity, Gender and the Sociology of Disasters*. London: Routledge.

Peacock, W. G. and A. K. Ragsdale. 1997. "Social systems, ecological networks and disasters: Toward a socio-political ecology of disasters." Pp. 20–35 in W. G. Peacock, W. G., B. H. Morrow, and H. Gladwin (eds.) *Hurricane Andrew: Ethnicity, Gender and the Sociology of Disasters*. London: Routledge.

Perrow, C. 1984. *Normal Accidents: Living With High-Risk Technologies*. New York: Basic Books.

Perrow, C. 1994. "The limits of safety: The enhancement of a theory of accidents." *Journal of Contingencies and Crisis Management* 2: 212–220.

Perry, J. B., R. Hawkins, and D. M. Neal. 1983. "Giving and receiving aid." *International Journal of Mass Emergencies and Disasters* 1: 171–187.

Perry, R. W. 1979a. "Evacuation decision making in natural disasters." *Mass Emergencies* 4: 25–38.

Perry. R. W. 1979b. "Incentives for evacuation in natural disaster." *Journal of the American Planning Association* 45: 440–447.

Perry, R. W. 1982. *The Social Psychology of Civil Defense.* Lexington, MA: DC Heath.

Perry, R. W. 1985. *Comprehensive Emergency Management: Evacuating Threatened Populations.* Greenwich, CT: JAI Press.

Perry, R. W. 1987. "Disaster preparedness and response among minority citizens." Pp. 135–151 in R. R. Dynes, B. De Marchi, and C. Pelanda (eds.) *Sociology of Disasters: Contribution of Sociology to Disaster Research.* Milan: Franco Angeli.

Perry, R. W. 1990. "Volcanic hazard perceptions at Mt. Shasta." *The Environmental Professional* 12: 312–318.

Perry, R. W. 1991. "Managing disaster response operations." Pp. 201–223 in T. E. Drabek and G. J. Hoetmer (eds.) *Emergency Management: Principles and Practice for Local Government.* Washington, DC: International City and County Management Association.

Perry, R. W. and M. R. Greene. 1982. "The role of ethnicity in emergency processes." *Sociological Inquiry* 52: 306–334.

Perry, R. W. and M. R. Greene, 1983. *Citizen Response to Volcanic Eruptions: The Case of Mt. St. Helens.* New York: Irvington Publishers.

Perry. R. W. and H. Hirose. 1991. *Volcano Management in the United States and Japan.* Greenwich, CT: JAI Press.

Perry, R. W. and M. K. Lindell. 1978. "The psychological consequences of natural disasters." *Mass Emergencies* 3: 105–117.

Perry, R. W. and M. K. Lindell. 1989. "Communicating threat information for volcanic hazards." Pp. 47–62 in L. M. Walters, L. Wilkins, and T. Walters (eds.) *Bad Tidings: Communication and Catastrophe.* Hillsdale, NJ: Lawrence Erlbaum Associates.

Perry. R. W. and M. K. Lindell. 1990a. *Living With Mt. St. Helens: Human Adjustment to Volcano Hazards.* Pullman, WA: Washington State University Press.

Perry, R. W. and M. K. Lindell. 1990b. "Citizen knowledge of volcano threats at Mt. St. Helens." *The Environmental Professional* 12: 45–51.

Perry, R. W. and M. K. Lindell. 1990c. "Predicting long term adjustment to volcano hazard." *International Journal of Mass Emergencies and Disasters* 8: 117–136.

Perry, R. W. and M. K. Lindell. 1991. "The effects of ethnicity on evacuation." *International Journal of Mass Emergencies and Disasters* 9: 47–68.

Perry, R. W. and M. K. Lindell. 1997. "Earthquake planning for governmental continuity." *Environmental Management* 21: 89–96.

Perry, R. W. and A. Mushkatel. 1984. *Disaster Management: Warning Response and Community Relocation.* Westport, CT: Quorum Books.

Perry, R. W. and A. Mushkatel. 1986. *Minority Citizens in Disasters.* Athens: University of Georgia Press.

Perry, R. W. and L. S. Nelson. 1991. "Ethnicity and hazard information dissemination." *Environmental Management* 15: 581–587.

Perry. R. W., D. F. Gillespie, and D. S. Mileti. 1974. "System stress and the persistence of emergent organizations." *Sociological Inquiry* 44: 111–119.

Perry, R. W., M. K. Lindell, and M. R. Greene. 1981. *Evacuation Planning in Emergency Management.* Lexington, MA: Lexington Books.

Perry, R. W., M. K. Lindell, and M. R. Greene. 1982. "Crisis communications: Ethnic differentials in interpreting and responding to disaster warnings." *Social Behavior and Personality* 10: 97–104.

Petak, W. J. (ed.) 1985. Special Issue: Emergency Management: A Challenge for Public Administration. Special Issue, *Public Administration Review* 45: 1–172.

Phifer, J. F. and F. H. Norris. 1989. "Psychological symptoms in older adults following natural disaster: Nature, timing, duration and course." *Journal of Gerontology* 44: S207–S217.

Phifer, J. F., K. Z. Kaniasty, and F. H. Norris. 1988. "The impact of natural disaster on the health of older adults: A multiwave prospective study." *Journal of Health and Social Behavior* 29: 65–78.

Phillips, B. D. 1990. "Gender as a variable in emergency response." Pp. 84–90 in R. Bolin (ed.) *The Loma Prieta Earthquake: Studies of Short-Term Impacts.* Boulder, CO: University of Colorado, Institute of Behavioral Science, Program on Environment and Behavior.

Phillips, B. D. 1991. "Post-Disaster Sheltering and Housing of Minority Groups." Final project report. Denton, TX: Department of Sociology and Social Work, Texas Woman's University.

Phillips, B. D. 1992. "Planning for the unexpected: Evacuation in a chemical explosion." *Disaster Management* 4: 103–108.

Phillips. B. D. 1993. "Cultural diversity in disasters: Sheltering, housing, and long term recovery." *International Journal of Mass Emergencies and Disasters* 11: 99–110.

Phillips, B. D. 1996. "Creating, sustaining, and losing place: Homelessness in the context of disaster." *Humanity and Society* 20: 94–101.

Phillips. B. D. 1998. "Sheltering and housing of low-income and minority groups in Santa Cruz County after the Loma Prieta earthquake." Pp. 17–28 in J. M. Nigg (ed.) *The Loma Prieta, California, Earthquake of October 17, 1989—Recovery, Mitigation, and Reconstruction.* Washington, DC: U.S. Geological Survey Professional Paper 1553-D.

Phillips, B. D., L. Garza, and D. N. Neal. 1994. "Intergroup relations in disasters: Service delivery barriers after Hurricane Andrew." *Journal of Intergroup Relations* 21: 18–27.

Picou, J. S., D. A. Gill, C. L. Dyer, and E. W. Curry. 1992. "Stress and disruption in an Alaskan fishing community: Initial and continuing impacts of the Exxon Valdez oil spill." *Industrial Crisis Quarterly* 6: 235–257.

Pijawka, K. D., B. A. Cuthbertson, and R. S. Olson. 1987–1988. "Coping with extreme hazard events: Emerging themes in natural and technological disaster research." *Omega* 18: 281–297.

Porfiriev, B. N. 1998. "Issues in the definition and delineation of disasters and disaster areas." Pp. 31–55 in E. L. Quarantelli (ed.) *What Is a Disaster? Perspectives on the Question.* London: Routledge.

Prince, S. H. 1920. "Catastrophe and Social Change; Based Upon a Sociological Study of the Halifax Disaster." Doctoral thesis. New York: Columbia University, Department of Political Science.

Quarantelli, E. L. 1954. "The nature and conditions of panic." *American Journal of Sociology* 60: 267–275.

Quarantelli, E. L. 1966. "Organizations under stress." Pp. 3–19 in Robert Bricston (ed.) *Symposium on Emergency Operations.* Santa Monica, CA: Systems Development Corporation.

Quarantelli, E. L. 1977. "Panic behavior: some empirical observations." Pp. 336–350 in *Human Response to Tall Buildings.* Stroudsburg, PA: Dowden, Hutchinson, and Ross, Inc.

Quarantelli, E. L. 1978. "Uses and problems of local EOCs in disasters." Newark, DE: University of Delaware, Disaster Research Center. Preliminary Paper no. 53.

Quarantelli, E. L. 1981a. *Sociobehavioral Responses to Chemical Hazards: Preparations for and Responses to Acute Chemical Emergencies at the Local Community Level.* Newark, DE: University of Delaware, Disaster Research Center. Final Project Report No. 28.

Quarantelli, E. L. 1981b. "The command post point of view in local mass communication systems." *Communications: International Journal of Mass Communication Research* 7: 57–73.

Quarantelli, E. L. 1982a. "What is a disaster? An agent specific or an all disaster spectrum approach to socio-behavioral aspects of earthquakes." Pp. 453–478 in B. G. Jones and M. Tomazevic (eds.) *Social and Economic Aspects of Earthquakes*. Ithaca, NY: Program in Urban and Regional Studies, Cornell University.

Quarantelli, E. L. 1982b. *Sheltering and Housing After Major Community Disasters: Case Studies and General Observations*. Newark, DE: Disaster Research Center. Final Project Report No. 29.

Quarantelli, E. L. 1982c. "General and particular observations on sheltering and housing after American disasters." *Disasters* 6: 277–281.

Quarantelli, E. L. 1982d. "Ten research-derived principles of disaster planning." *Disaster Management* 2: 23–25.

Quarantelli, E. L. 1983. *Delivery of Emergency Medical Services in Disasters*. New York: Irvington.

Quarantelli, E. L. 1984. "Chemical disaster preparedness at the local community level." *Journal of Hazardous Materials* 8: 239–249.

Quarantelli, E. L. 1985. "What is Disaster? The need for clarification in definition and conceptualization in research." Pp. 41–73 in B. J. Sowder (ed.) *Disasters and Mental Health: Selected Contemporary Perspectives*. Rockville, MD. National Institute of Mental Health.

Quarantelli, E. L. 1987. "Disaster studies: An analysis of the social historical factors affecting the development of research in the area." *International Journal of Mass Emergencies and Disasters* 5: 285–310.

Quarantelli, E. L. 1988. "Assessing disaster preparedness planning." *Regional Development Dialogue* 9: 48–69.

Quarantelli, E. L. 1989a. "Conceptualizing disasters from a sociological perspective." *International Journal of Mass Emergencies and Disasters* 7: 243–251.

Quarantelli, E. L. 1989b. "The social science study of disasters and mass communication." Pp. 1–19 in L. M. Walters, L. Wilkins, and T. Walters (eds.) *Bad Tidings: Communication and Catastrophe*. Hillsdale, NJ: Lawrence Erlbaum Associates.

Quarantelli, E. L. 1991. "The mass media in disasters in the United States." Pp. 348–353 in *Proceedings of the IDNDR International Conference 1990*. Tokyo: Japanese Government Headquarters for the IDNDR.

Quarantelli, E. L. 1992. "The case for a generic rather than agent specific approach to disasters." *Disaster Management* 2: 191–196.

Quarantelli, E. L. 1993. "Converting disaster scholarship into effective disaster planning and managing: Possibilities and limitations." *International Journal of Mass Emergencies and Disasters* 11: 15–39.

Quarantelli, E. L. 1997. "Problematical aspects of the information/communication revolution for disaster planning and research: Ten non-technical issues and questions." *Disaster Prevention and Management* 6: 94–106.

Quarantelli, E. L. 1998a. *What Is a Disaster? Perspectives on the Question*. London: Routledge.

Quarantelli, E. L. 1998b. "The computer based information/communication revolution: A dozen problematical issues and questions they raise for disaster planning and managing." Newark, DE: University of Delaware, Disaster Research Center.

Quarantelli, E. L. 1998c. "The proposed establishment of global disaster information network (GDIN): The social dimensions involved." Newark, DE: University of Delaware, Disaster Research Center.

Quarantelli, E. L. and R. R. Dynes. 1970a. "Property norms and looting." *Phylon* 31: 168–182.

Quarantelli, E. L. and R. R. Dynes (eds.) 1970b. Special Issue: Organizational and Group Behavior in Disasters. *American Behavioral Scientist* 13: 323–480.

Quarantelli, E. L. and R. R. Dynes. 1972. "When disaster strikes (it isn't much like what you've heard and read about)." *Psychology Today* 5: 66–70.

Quarantelli, E. L. and R. R. Dynes. 1976. "Community conflict: Its absence and its presence in natural disasters." *Mass Emergencies* 1: 139–152.

Quarantelli, E. L. and R. R. Dynes. 1977. "Response to Social Crisis and Disaster." *Annual Review of Sociology* 3: 23–49.

Quarantelli, E. L., C. Lawrence, K. J. Tierney, and Q. T. Johnson. 1979. "Initial findings from a study of socio-behavioral preparations and planning for acute chemical hazard disasters." *Journal of Hazardous Materials* 3: 79–90.

Quarantelli, E. L., with K. E. Green, E. Ireland, S. McCabe, and D. M. Neal. 1983. *Emergent Citizen Groups in Disaster Preparedness and Recovery Activities: An Interim Report.* Newark, DE: University of Delaware, Disaster Research Center.

Rahimi, M. and G. Azevedo. 1993. "Building content hazards and behavior of mobility-restricted residents." Pp. 51–62 in P. A. Bolton (ed.) *The Loma Prieta, California, Earthquake of October 17, 1989—Public Response.* U.S. Geological Survey Professional Paper 1553-B. Washington, DC: U.S. Government Printing Office.

Riad, J. K., F. H. Norris, and R. B. Ruback. 1999. "Predicting evacuation in two major disasters: Risk perception, social influence, and access to resources." *Journal of Applied Social Psychology* 29: 918–934.

Roberts, K. H. 1989. "New challenges in organizational research: High-reliability organizations." *Industrial Crisis Quarterly* 3:111–125.

Roberts, K. H., D. M. Rousseau, and T. R. La Porte. 1993. "The culture of high reliability: quantitative and qualitative assessment about nuclear powered aircraft carriers." *High Technology Management Research* 5: 141–161.

Rogers, G. O. and J. H. Sorensen. 1991. "Adoption of emergency planning practices for chemical hazards in the United States." *Journal of Hazardous Materials* 27: 3–26.

Rogers, E. M. and R. Sood. 1981. "Mass media operations in a quick-onset natural disaster: Hurricane David in Dominica." Boulder, CO: University of Colorado, Institute of Behavioral Science. Working Paper No. 41.

Rogers. E. M., J. Berndt, J. Harris, and J. Minzer. 1990. *Mass Media Coverage of the 1989 Loma Prieta Earthquake: Estimating the Severity of a Disaster.* Los Angeles, CA: University of Southern California, Annenberg School for Communication.

Rosen, R. 1994. "Who Gets Polluted?" *Dissent,* Spring: 223–230.

Ross, G. A. 1980. "The emergence of organization sets in three ecumenical recovery organizations: An empirical and theoretical exploration." *Human Relations* 33: 23–29.

Rossi, P. H., J. D. Wright, and E. Weber-Burdin. 1982. *Natural Hazards and Public Choice: The State and Local Politics of Hazard Mitigation.* New York: Academic Press.

Ruch, C. 1991. "Behavioral and logistical feasibility of vertical evacuation." Pp. 7–22 in C. Ruch, H. C. Miller, M. Haflich, N. M. Farber, P. R. Berke, and N. Stubbs (eds.). *The Feasibiity of Vertical Evacuation.* Boulder, CO: University of Colorado, Institute of Behavioral Science, Monograph No. 52.

Ruch, C., H. C. Miller, M. Haflich, N. M. Farber, P. R. Berke, and N. Stubbs (eds.) 1991. *The Feasibiity of Vertical Evacuation.* Boulder, CO: University of Colorado, Institute of Behavioral Science, Monograph No. 52.

Ruggles, S. 1994. "The origins of African-American family structure." *American Sociological Review* 59: 136–151.

Russell, G. W. and R. K. Mentzel. 1990. "Sympathy and altruism in response to disaster." *Journal of Social Psychology* 130: 309–317.

Russell, L. A., J. D. Goltz, and L. B. Bourque. 1995. "Preparedness and hazard mitigation activities before and after two earthquakes." *Environment and Behavior* 27: 744–770.

Saarinen, T. and J. Sell. 1985. *Warning and Response to the Mt. St. Helens Eruption.* Albany, NY: State University of New York Press.

Sagan, S. D. 1993. *The Limits of Safety: Organizations, Accidents, and Nuclear Weapons.* Princeton, NJ: Princeton University Press.

Sarbin, T. R. and J. I. Kitsuse (eds.). 1994. *Constructing the Social.* Thousand Oaks, CA: Sage Publications.

Scanlon, J. 1994. "The role of EOCs in emergency management: A comparison of American and Canadian experience." *International Journal of Mass Emergencies and Disasters* 12: 51–75.

Scanlon, J. 1997. "Human behaviour in disaster: The relevance of gender." *Australian Journal of Emergency Management* 11: 2–7.

Scanlon, J., S. Alldred, A. Farrell, and A. Prawzick. 1985. "Coping with the media in disasters: Some predictable problems." *Public Administration Review* 45: 123–133.

Schnaiberg, A. 1980. *The Environment: From Surplus to Scarcity.* New York: Oxford University Press.

Schnaiberg, A. and K. A. Gould. 1994. *Environment and Society: The Enduring Conflict.* New York: St. Martin's Press.

Schneider, S. K. 1995. *Flirting with Disaster: Public Management in Crisis Situations.* Armonk, NY: M. E. Sharpe.

Schroeder, R. A. 1987. "Gender Vulnerability to Drought: A Case Study of the Hausa Social Environment." Boulder, CO: University of Colorado, Institute of Behavioral Science, Program on Environment and Behavior. Working Paper No. 80.

Seitz, S. T. and M. Davis. 1984. "The political matrix of natural disasters: Africa and Latin America." *International Journal of Mass Emergencies and Disasters* 2: 231–250.

Sherman, S. J. and E. Corty. 1984. "Cognitive heuristics." Pp. 189–286 in R. S. Wyer, Jr. and T. K. Srull (eds.) *Handbook of Social Cognition,* Vol. 1. Hillsdale, NJ: Erlbaum.

Shihadeh, E. S. and D. J. Steffensmeier. 1994. "Economic inequality, family disruption, and urban black violence: Cities as units of stratification and social control." *Social Forces* 73: 729–751.

Showalter, P. S. 1993. "Prognostication of doom: An earthquake prediction's effect on four small communities." *International Journal of Mass Emergencies and Disasters* 11: 297–292.

Shrestha, L. B. 1997. *Racial Differences in Life Expectancy Among Elderly African Americans and Whites.* New York: Garland Publishing.

Shrivastava, P. 1987. *Bhopal: Anatomy of a Crisis.* Cambridge, MA: Ballinger Publishing Company.

Sillar, W. 1975. "Planning for disasters." *Long-Range Planning* 8: 2–7.

Simile, C. 1995. "Disaster Settings and Mobilization for Contentious Collective Action: Case Studies of Hurricane Hugo and the Loma Prieta Earthquake." Doctoral dissertation. Newark, DE: Department of Sociology and Criminal Justice, University of Delaware.

Simpson, R. H. 1980. "Will coastal residents reach safe shelter in time?" Pp. 25–37 in E. J. Baker (ed.) *Hurricanes and Coastal Storms*. Tallahassee, FL: Florida State University Sea Grant Program.

Sims, J. and D. Bauman. 1972. "The tornado threat: Coping styles of the North and South." *Science* 176: 1386–1392.

Singer, T. 1982. "An introduction to disaster." *Aviation, Space, and Environmental Medicine* 53: 245–250.

Slovic, P., H. Kunreuther, and G. F. White. 1974. "Decision processes, rationality and adjustments to natural hazards." Pp, 187–204 in G. F. White (ed.) *Natural Hazards: Local, National, Global*. New York: Oxford University Press.

Smith, C. 1992. *Media and Apocalypse: News Coverage of the Yellowstone Forest Fires, Exxon Valdez Oil Spill, and Loma Prieta Earthquake*. Westport, CT: Greenwood Press.

Smith, E., L. N. Robins, T. R. Przybeck, E. Goldring, and S. D. Solomon. 1986. "Psychosocial consequences of a disaster." Pp. 49–76 in J. H. Shore (ed.) *Disaster Stress Studies: New Methods and Findings*. Washington, DC: American Psychiatric Press.

Solyst, J. and M. St. Amand. 1991. *Emergency Planning and Community Right to Know Act: A Status of State Actions—1991*. Washington, DC: National Governors' Association.

Sorensen, J. H. 1993. "Warning systems and public warning response." Paper presented at the workshop on Socioeconomic Aspects of Disaster in Central America, San Jose, Costa Rica.

Sorensen, J. H. and B. Richardson. 1984. "Risk and uncertainty as determinants of human response in emergencies: Evacuation at TMI reexamined." Proceedings of the annual meeting of the Society for Risk Analysis, Knoxville, TN.

Sorensen, J. H. and G. F. White. 1980. "Natural hazards: A cross-cultural perspective." Pp. 279–318 in I. Altman, A. Rapaport, and J. F. Wohlwill (eds.) *Human Behavior and the Environment: Advances in Theory and Research*. New York: Plenum Press.

Sorensen, J. H. and D. S. Mileti. 1987. "Programs that encourage the adoption of precautions against natural hazards: Reviews and evaluation." Pp. 208–230 in N. Weinstein (ed.) *Taking Care: Why People Take Precautions*. New York: Cambridge University Press.

Sorensen, J. H. and D. S. Mileti. 1989. "Warning systems for nuclear power plant emergencies." *Nuclear Safety* 30: 358–370.

Sorensen, J. H. and G. O. Rogers. 1988. "Local preparedness for chemical accidents: A survey of U.S. communities." *Industrial Crisis Quarterly* 2: 89–108.

Sorensen, J. H., D. S. Mileti, and E. Copenhaver. 1985. "Inter and intraorganizational cohesion in emergencies." *International Journal of Mass Emergencies and Disasters* 3: 27–52.

Sorensen, J. H., B. N. Vogt, and D. S. Mileti. 1987. *Evacuation: An Assessment of Planning and Research*. Washington, DC: Federal Emergency Management Agency.

Spangle, W. E. (ed.) 1987. *Pre-Earthquake Planning for Post-Earthquake Rebuilding*. Pasadena, CA: Governor's Office of Emergency Services, Southern California Earthquake Preparedness Project.

Spector, M. and J. I. Kitsuse. 1987. *Constructing Social Problems*. New York: Aldine de Gruyter.

Stallings, R. A. 1978. "The structural patterns of four types of organizations in disaster." Pp. 87–103 in E. L. Quarantelli (ed.) *Disasters: Theory and Research*. Thousand Oaks, CA: Sage Publications.

Stallings, R. A. 1984. "Evacuation behavior at Three Mile Island." *International Journal of Mass Emergencies and Disasters* 2: 11–26.

Stallings, R. A. 1988. "Conflict in natural disasters: A codification of consensus and conflict theories." *Social Science Quarterly* 69: 569–586.

Stallings, R. A. 1989. "Volunteerism inside complex organizations: Off-duty hospital personnel in a disaster." *Nonprofit and Voluntary Sector Quarterly* 18: 133–145.

Stallings, R. A. 1991. "Disasters as social problems? A dissenting view." *International Journal of Mass Emergencies and Disasters* 9: 69–74.

Stallings, R. A. 1995. *Promoting Risk: Constructing the Earthquake Threat.* New York: Aldine de Gruyter.

Stallings, R. A. 1998. "Disaster and the theory of social order." Pp. 127–145 in E. L. Quarantelli (ed.) *What is a Disaster? Perspectives on the Question.* London: Routledge.

Stallings, R. A. and E. L. Quarantelli. 1985. "Emergent citizen groups and emergency management." *Public Administration Review* 45: 93–100.

Stallings, R. A. and C. B. Schepart. 1987. "Contrasting local government responses to a tornado disaster in two communities." *International Journal of Mass Emergencies and Disasters* 5: 265–284.

Staples, R. 1976. *Introduction to Black Sociology.* New York: McGraw Hill.

Sugiman, T., T. Atsumi, M. Nagata, and T. Watanabe. 1996. "The process of organizing emergency shelters after the 1995 Hanshin earthquake." Pp. 367–374 in *Proceedings of the International Conference on Water Resources and Environmental Research* Vol. II, Kyoto, Japan, October 29–31.

Sullivan, R., D. A. Mustart, and J. S. Galehouse. 1977. "Living in earthquake country." *California Geology* 30: 3–8.

Susman, P., P. O'Keefe, and B. Wisner. 1983. "Global disasters, a radical interpretation." Pp. 263–283 in K. Hewitt (ed.) *Interpretations of Calamity.* Boston: Allen and Unwin.

Sylves, R. T. 1984. "Nuclear power emergency planning: Politics of the task." *International Journal of Mass Emergencies and Disasters* 2: 185–196.

Sylves, R. T. 1991. "Adopting integrated emergency management in the United States: Political and organizational challenges." *International Journal of Mass Emergencies and Disasters* 9: 413–424.

Taylor, V. A. 1976. "The Delivery of Mental Health Services in the Xenia Tornado: A Collective Behavior Analysis of an Emergent System Response." Doctoral dissertation. Columbus, OH: Department of Sociology, Ohio State University.

Taylor, V. A. 1977. "Good news about disasters." *Psychology Today,* October: 93–96.

Tierney, K. J. 1985a. "Emergency medical preparedness and response to disasters: The need for interorganizational coordination." *Public Administration Review* 45: 77–84.

Tierney, K. J. 1985b. *Report on the Coalinga Earthquake of May 2, 1983.* Sacramento, CA: California Seismic Safety Commission.

Tierney, K. J. 1988. "Social aspects of the Whittier Narrows earthquake." *Earthquake Spectra* 4: 11–23.

Tierney, K. J. 1989. "Improving theory and research on hazard mitigation: Political economy and organizational perspectives." *International Journal of Mass Emergencies and Disasters* 7: 367–396.

Tierney, K. J. 1992. "Socio-economic aspects of hazard mitigation." Newark, DE: University of Delaware, Disaster Research Center. Preliminary Paper No.190.

Tierney, K. J. 1993. *Project Summary: Disaster Analysis: Delivery of Emergency Medical Services in Disasters.* Newark, DE: University of Delaware, Disaster Research Center.

Tierney, K. J. 1994. "Emergency preparedness and response." Pp. 105–128 in *Practical Lessons from the Loma Prieta Earthquake.* Washington, DC: National Academy Press.

Tierney, K. J. 1998. "The field turns fifty: Social change and the practice of disaster field work." Newark, DE: University of Delaware, Disaster Research Center. Preliminary Paper No. 273.

Tierney, K. J. 1999. "Toward a critical sociology of risk." *Sociological Forum* 14: 215–242.

Tierney, K. J. and J. D. Goltz. 1997. "Emergency response: Lessons learned from the Kobe earthquake." Newark, DE: University of Delaware, Disaster Research Center. Preliminary Paper No. 260.

Tierney, K. J., W. J. Petak, and H. Hahn. 1988. *Disabled Persons and Earthquake Hazards.* Boulder, CO: University of Colorado, Institute of Behavioral Science, Program on Environment and Behavior.

Tinker, J. 1984. "Are natural disasters natural?" *Socialist Review* 14: 7–25.

Tuchman, G. 1978. *Making News: A Study in the Construction of Reality.* New York: Free Press.

Turner, R. H. and L. M. Killian. 1987. *Collective Behavior.* Englewood Cliffs, NJ: Prentice-Hall.

Turner, R. H., J. M. Nigg, and D. Heller-Paz. 1986. *Waiting for Disaster: Earthquake Watch in California.* Berkeley: University of California Press.

Turner, R. H., J. M. Nigg, D. Heller-Paz, and B. Young. 1981. *Community Response to the Earthquake Threat in Southern California.* Los Angeles: University of California, Los Angeles, Institute for Social Science Research.

Tversky, A. and D. Kahneman. 1973. "Availability: A heuristic for judging frequency and probability." *Cognitive Psychology* 4: 207–232.

Tversky, A. and D. Kahneman. 1981. "The framing of decisions and the psychology of choice." *Science* 211: 453–458.

U.S. Department of Commerce. 1995. *A Summary of Natural Hazard Fatalities for 1994 in the United States.* Silver Spring, MD: National Weather Service, Dept. of Meteorology.

U.S. General Accounting Office. 1991. *Disaster Assistance: Federal, State, and Local Responses to Natural Disasters Need Improvement.* Washington, DC: General Accounting Office.

Urbanik, T. 1994. "State of the art in evacuation time estimate studies for nuclear power plants." *International Journal of Mass Emergencies and Disasters* 12: 327–343.

Urbanik, T. H., A. C. Desrosiers, M. K. Lindell, and C. R. Schuller. 1980. *Analysis of Techniques for Estimating Evacuation Times for Emergency Planning Zones.* Washington, DC: U.S. Nuclear Regulatory Commission. NUREG/CR-1745.

Vallance, T. and A. D'Augelli. 1982. "The helpers community: Characteristics of natural helpers." *American Journal of Community Psychology* 10: 197–205.

Vaughan, D. 1996. *The Challenger Launch Decision: Risky Technology, Culture, and Deviance at NASA.* Chicago: University of Chicago Press.

Vaughan, D. 1999. "The dark side of organizations: Mistake, misconduct, and disaster." *Annual Review of Sociology* 25: 271–305.

Vogt, B. M. 1991. "Issues in nursing home evacuations." *International Journal of Mass Emergencies and Disasters* 9: 247–265.

Vogt, B. M. and J. H. Sorensen. 1987. *Evacuation in Emergencies: An Annotated Guide to Research.* Oak Ridge, TN: Oak Ridge National Laboratory. Report No. ORNL/TM-10277.

Wallace, A. F. C. 1957. "Mazeway distintegration." *Human Organization* 16: 23–27.

Warheit, G. J. and J. Waxman. 1973. "Operational and organizational adaptations of fire departments to civil disturbances." *American Behavioral Scientist* 16: 343–355.

Warren, K. J. (ed.) with B. Wells-Howe. 1994. *Ecological Feminism.* London: Routledge.

Watson, E. and A. Collins. 1982. "Natural helping networks in alleviating family stress." *The Annals of the American Academy of Political and Social Science* 461: 102–112.

Waugh, W. L. Jr. 1988. "Current policy and implementation issues in disaster prepared-ness." Pp. 111–125 in L. K. Comfort (ed.) *Managing Disaster: Strategies and Policy Perspectives.* Durham, NC: Duke University Press.

Waugh, W. L., Jr. and R. T. Sylves. 1996. "The intergovernmental relations of emergency management." Pp. 46–68 in R. T. Sylves and W. L. Waugh, Jr. (eds.) *Disaster Man-agement in the U.S. and Canada,* 2nd edition. Springfield, IL: Charles C Thomas.

Webb, G. R., K. J. Tierney, and J. M. Dahlhamer. 2000. "Businesses and disasters: Empiri-cal patterns and unanswered questions." *Natural Hazards Review* 1, 2: 83–90.

Weller, J. M. and E. L. Quarantellli. 1973. "Neglected characteristics of collective behav-ior." *American Journal of Sociology* 79: 665–685.

Weller, J. M. and D. E. Wenger. 1973. "Disaster subcultures: The cultural residues of community disasters." Newark, DE: University of Delaware, Disaster Research Cen-ter. Preliminary Paper No. 9.

Wenger, D. E. 1972. "DRC studies of community functioning." Pp. 29–72 in *Proceedings of the Japan-United States Disaster Research Seminar.* Newark, DE: University of Delaware, Disaster Research Center.

Wenger, D. E. 1973. "The reluctant army: The functioning of police departments during civil disturbances." *American Behavioral Scientist* 16: 326–342.

Wenger, D. E. 1978. "Community response to disaster: Functional and structural alter-ations." Pp. 18–47 in E. L. Quarantelli (ed.) *Disasters: Theory and Research.* Thou-sand Oaks, CA: Sage Publications.

Wenger, D. E. 1985. "Mass Media and Disasters." Newark, DE: Disaster Research Center. Preliminary Paper No. 98.

Wenger, D. E. 1987. "Collective behavior and disaster research." Pp. 213–237 in R. R. Dynes, B. De Marchi, and C. Pelanda (eds.) *Sociology of Disasters: Contribution of Sociology to Disaster Research.* Milan: Franco Angeli.

Wenger, D. E. 1991. "Emergent and volunteer behavior during disaster: Research findings and planning implications." College Station, TX: Texas A&M University, Hazard Reduction and Recovery Center.

Wenger, D. E. and T. F. James. 1994. "The convergence of volunteers in a consensus crisis: The case of the 1985 Mexico City earthquake." Pp. 229–243 in R. R. Dynes and K. J. Tierney (eds.) *Disasters, Collective Behavior, and Social Organization.* Newark, DE: University of Delaware Press.

Wenger, D. and E. L. Quarantelli, 1989. *Local Mass Media Operations, Problems, and Products in Disasters.* Newark, DE: Disaster Research Center, University of Dela-ware. Report Series No. 19.

Wenger, D., E. L. Quarantelli, and R. R. Dynes. 1986. "Disaster Analysis: Local Emer-gency Management Offices and Arrangements." Newark, DE: University of Dela-ware, Disaster Research Center. Final Project Report No. 34.

Wenger, D., E. L. Quarantelli, and R. R. Dynes. 1989. "Disaster Analysis: Police and Fire Departments." Newark, DE: University of Delaware, Disaster Research Center. Final Project Report No. 37.

Wenger, D. E., J. D. Dykes, T. D. Sebok, and J. L. Neff. 1975. "It's a matter of myths: An empirical examination of individual insight into disaster response." *Mass Emergen-cies* 1:33–46.

White, G. F. 1974. *Natural Hazards: Local, National, Global.* New York: Oxford Univer-sity Press.

White, G. F. and J. E. Haas. 1975. *Assessment of Research on Natural Hazards.* Cam-bridge, MA: MIT Press.

Whitney, D. J. and M. K. Lindell. 2000. "Member commitment and participation in Local Emergency Planning Committees." *Policy Studies Journal* 28: 467–484.

Whitney, D. J. A. Dickerson, and M. K. Lindell. 1999. *Non-structural Seismic Preparedness of Southern California Hospitals.* College Station, TX: Texas A&M University, Hazard Reduction and Recovery Center.

Wijkman, A. and L. Timberlake. 1988. *Natural Disasters: Acts of God or Acts of Man?* Philadelphia: New Society Publishers.

Wilkins, L. 1987. *Shared Vulnerability: The Mass Media and American Perception of the Bhopal Disaster.* Westport, CT: Greenwood Press.

Wilkins, L. 1989. "Bhopal: The politics of mediated risk." Pp. 21–34 in L. Masel Walters, L. Wilkins, and T. Walters (eds.) *Bad Tidings: Communication and Catastrophe.* Hillsdale, NJ: Lawrence Erlbaum Associates.

Wilkins, L. 1996. "Living with the flood: Human and governmental responses to symbolic risk." Pp. 218–244 in S. A. Chagnon (ed.) *The Great Flood of 1993: Causes, Impacts, and Responses.* Boulder, CO: Westview Press.

Wilmer, H. 1958. "Toward a definition of the therapeutic community." *American Journal of Psychiatry* 114: 824–834.

Wolensky, R. P. and K. C. Wolensky. 1991. "American local government and the disaster management problem." *Local Government Studies*, March/April: 15–32.

Wolff, E. N. 1996. *Top Heavy: A Study of the Increasing Inequality of Wealth in America.* New York: Twentieth Century Fund Press.

Wood, P. G. 1980. "A survey of behaviour in fires." Pp. 83–96 in D. Canter (ed.) *Fires and Human Behaviour.* Chichester, UK: Wiley.

Wood, R. and A. Bandura. 1989. "Social cognitive theory of organizational management." *Academy of Management Review* 14: 361–383.

Yelvington, K. A. 1997. "Coping in a temporary way: The tent cities." Pp. 92–115 in W. G. Peacock, B. H. Morrow, and H. Gladwin (eds.) *Hurricane Andrew: Ethnicity, Gender and the Sociology of Disasters.* London: Routledge.

Young, C., S. Giles, and M. Plantz. 1982. "Natural helping networks." *American Journal of Community Psychology* 10: 457–469.

Zeigler, D. J. and J. H. Johnson. 1984. "Evacuation behavior in response to nuclear power plant accidents." *Professional Geographer* 36: 207–215.

Zeigler, D. J., S. D. Brunn, and J. H. Johnson. 1981. "Evacuation from a nuclear technological disaster." *Geographical Review* 71: 1–16.

Index